Soil and Land Capability for Agriculture

EASTERN SCOTLAND

By A. D. Walker, BSc, C. G. B. Campbell, BSc,
R. E. F. Heslop, BSc, J. H. Gauld, BSc, PhD,
D. Laing, BSc, C Chem., M.R.S.C., B. M. Shipley, BSc
and G. G. Wright, BSc
with a contribution by J. S. Robertson, BSc

The Macaulay Institute for Soil Research
Aberdeen 1982

Front cover. *The fertile vale of Strathmore from Aldbar Castle, near Brechin, looking north towards the Grampian Highlands. The flat, arable fields in the foreground and middle distance are on brown forest soils of map unit 41 (Balrownie Association) and the humus-iron podzols and brown forest soils of map unit 238 (Forfar Association), both Class 2; in the far distance, the undulating lowlands are Class 3.1 and 3.2. On the lower slopes and foothills are humus-iron podzols of map unit 276 (Strathfinella Association) and map unit 498 (Strichen Association), mainly Classes 4 and 5. Peaty podzols of map unit 499 (Strichen Association, Class 6.2) dominate the upper slopes and much of the plateau.* Aerofilms.

ISBN 0 7084 0223 2

PRINTED IN GREAT BRITAIN
AT THE UNIVERSITY PRESS
ABERDEEN

Contents

LIST OF PLATES

LIST OF FIGURES

CONTENTS

LIST OF TABLES

Preface

Extensive soil surveys had been undertaken in the lowlands and foothills of Eastern Scotland prior to 1978. The derived maps, published at a scale of 1:63 360, included the following areas: Stirling, Kinross and Elie, Perth and Arbroath, Forfar, Banchory and Stonehaven, Inverurie, Aberdeen, The Black Isle, Nairn and Cromarty, Peterhead and Fraserburgh, Cromarty and Invergordon, Elgin and Banff.

Upon approval of the proposal for the soil survey of the remaining parts of Scotland at a scale of 1:250 000, mapping was carried out in 1978, 1979 and 1980 by J. S. Bell, T. W. M. Brown, C. G. B. Campbell, R. E. F. Heslop, J. H. Gauld, D. Laing, B. M. Shipley, A. D. Walker and G. G. Wright. The overlap with Northern Scotland region (Sheet 3) was surveyed by C. G. B. Campbell, D. W. Futty, R. E. F. Heslop, A. J. Nolan, W. Towers, A. D. Walker and G. G. Wright. The western fringe of Sheet 5 lying in the overlap with the Western Scotland region (Sheet 4) was surveyed by J. S. Bell, T. W. M. Brown, J. H. Gauld, B. M. Shipley, A. D. Walker and G. G. Wright. The overlap with the Southern Scotland region (Sheets 6 and 7) was surveyed by J. S. Bell, D. Laing and B. M. Shipley. The areas of responsibility for mapping are shown in Fig. 1. Compilation of the soil maps was carried out during 1981 based on a National Soil Map Legend devised by B. M. Shipley. Analytical data quoted in the text were produced at the Macaulay Institute for Soil Research, Aberdeen, mostly in the Department of Mineral Soils. The vegetation assessments were carried out by field staff according to a system designed by E. L. Birse and J. S. Robertson; correlation was the responsibility of the latter who also wrote the account of the plant communities. The authors of the Eastern Scotland Handbook were C. G. B. Campbell, R. E. F. Heslop, J. H. Gauld, D. Laing, B. M. Shipley, A. D. Walker and G. G. Wright. The handbook has been compiled by A. D. Walker and edited by D. W. Futty.

The base map was compiled and drawn by the Soil Survey cartographic section using modified components from Ordnance Survey 1:250 000 scale topographic and administrative maps. The maps were drafted by W. S. Shirreffs and Miss P. R. Carnegie. The diagrams in this book were drawn by A. D. Moir and Mrs. R. M. J. Fulton.

Concurrently with the soil mapping, the staff of the Survey Department carried out assessments of land capability for agriculture using guidelines devised by Bibby, Douglas, Thomasson and Robertson (1982). Advisory groups were established to assist the surveyors in this task. They consisted of representatives of the Department of Agriculture and Fisheries for Scotland, the Scottish Agricultural Colleges and the National Farmers' Union of Scotland. In addition, consultation with the local offices of the various organizations was maintained. The committees proved lively forums for discussion and made valuable contributions to the interpretative maps. The responsibility for the maps, however, remains entirely with the Soil Survey of Scotland.

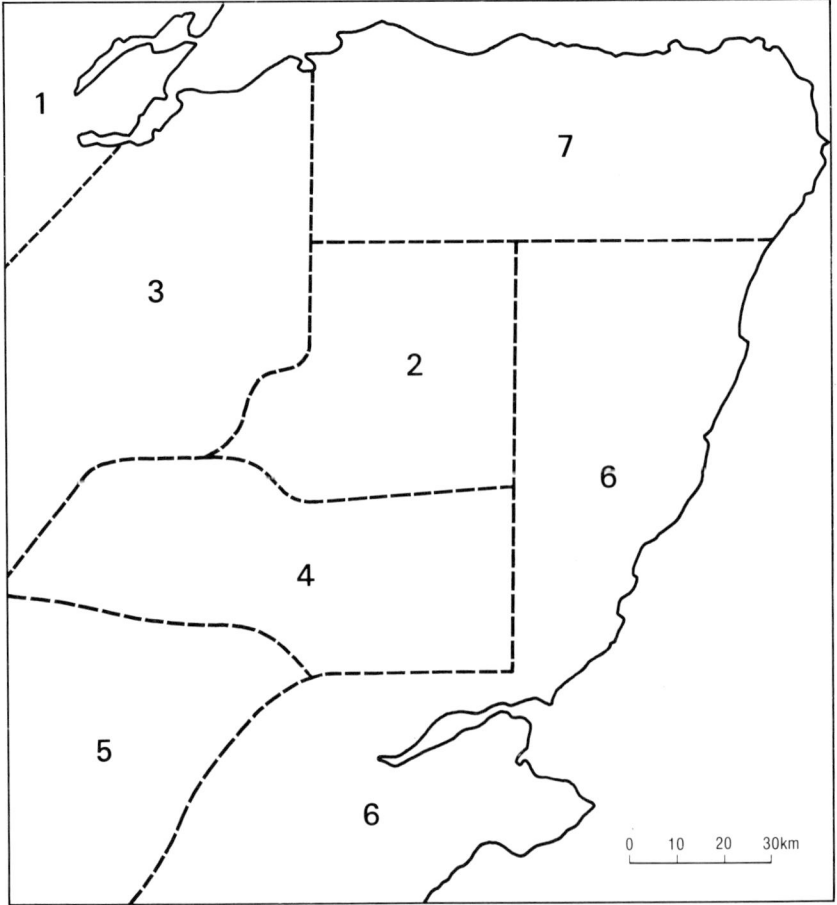

1   R. E. F. Heslop, D. W. Futty, C. G. B. Campbell, W. Towers and A. J. Nolan

2   R. E. F. Heslop and C. G. B. Campbell

3   A. D. Walker and G. G. Wright

4   J. H. Gauld and J. S. Bell

5   B. M. Shipley and T. W. M. Brown

6   D. Laing

7   D. Laing, B. M. Shipley, R. E. F. Heslop, C. G. B. Campbell and G. G. Wright

Figure 1. *Survey teams' map areas.*

The aerial photographs (scale c. 1:25 000) and copies of the field maps (scale 1:50 000) used in the survey may be inspected by prior arrangement with the Department of Soil Survey, The Macaulay Institute for Soil Research, Craigiebuckler, Aberdeen, AB9 2QJ.

ROBERT GRANT

Head of the Soil Survey of Scotland

Acknowledgements

The Department of Soil Survey wishes to thank the many landowners and farmers who willingly co-operated in the survey by allowing access to their land. The assistance of various agricultural organizations in the land capability assessments has already been acknowledged, but the Department would like to thank in particular, the following (listed in alphabetical order) for their valuable assistance and contributions to the advisory committees.

I. G. Alexander, G. Buchanan, C. G. Davidson, J. Davidson, F. M. B. Houston, W. McGregor, I. Mathieson, R. S. Patterson, W. C. Robbie and J. Valentine (Department of Agriculture and Fisheries for Scotland); A. J. M. Mackay, R. G. Tate, and A. J. Taylor, (East of Scotland College of Agriculture); H. Black, D. Findlay, J. W. Grant, A. Howie, I. Lumsden and J. Vallance (North of Scotland College of Agriculture); and R. F. Anderson, J. Boyne, M. K. Browne, N. J. Donaldson, D. Mackenzie, A. A. Meldrum, A. Pratt, A. M. Reid and Miss N. A. Wright (National Farmers' Union of Scotland).

In addition the Department would like to acknowledge the considerable assistance in the field afforded by M. K. Browne, (N. F. U.).

Photographs in the text are by members of the Soil Survey of Scotland, Aerofilms Ltd, the Institute of Geological Sciences, Cambridge University, Ordnance Survey, the Scottish Development Department, RAE Farnborough, A D S Macpherson and Aberdeen Journals.

1 Description of the Area

LOCATION AND EXTENT

Eastern Scotland (Sheet 5) covers a total land area of 24 544 square kilometres and spans an altitude range from sea level to 1309 metres on Ben Macdui, the second highest peak in the United Kingdom. Lying mainly between the Moray Firth and the Firth of Forth within the eastern watershed of the central mass of Scotland, the region (Plate I and Fig. 2) is dominated by the Grampian Highlands. These comprise three areas each with a characteristic landscape. In the North-East Grampian Highlands, the hills culminate in the vast plateaux of the Cairngorm Mountains. Though occasionally scalloped by enormous corries and deeply incised by glacial troughs, they are characterized by the smooth slopes of planation surfaces which descend in distinct steps to the Central Lowlands and the North-East Lowlands. By contrast, the South-West Grampian Highlands provide a rugged and highly dissected terrain. The narrow deep lochs, so typical of the area, occupy the rock basin outlets used by the ice which moved radially from the vast cauldron of Rannoch Moor during the glacial period. North of an intermediate zone around the Loch Laggan trough, lie the Monadhliath Mountains, the third element of the Grampian Highlands. Consisting of a sea of rolling, peat-covered, broad hills and plateaux, these mountains descend gently northwards from about 900 metres to merge with the 300-metre peneplain and the foothills fringing the Moray Firth Lowlands. Across the Great Glen lies the south-east margin of the Northern Highlands.

Bordering the highlands to the north, north-east and south-east, the lowlands form the main arable belt in Scotland; each has its own identity reflecting the different lithology and evolution. Partly as a consequence of the major rivers trending in a north-easterly or easterly direction, access to much of the mountainous hinterland is moderately good. Only in the Monadhliath and Cairngorm Mountains is there still a problem though the rapidly developing network of bulldozed hill-tracks is reducing the difficulty. During the last two decades, the spanning of all the major coastal re-entrants by new road-links has removed the final barriers to speedy and effective transportation across the lowlands. This combination of good climate, high agricultural productivity, ease of access and natural harbours has led to the historical development of many primary population centres along the coast.

1

Plate 1. *Landsat Grampian image, band 7. Taken from approximately 900 Km above the surface of the earth.* Image made by Space Department, RAE Farnbourgh.

In the foothills, stock-rearing, sport and afforestation are competing forms of land-use though an integrated approach is slowly becoming more widely practised. The once widespread Caledonian Forest has been reduced to a few remnants mainly in Speyside and Rannoch. Afforestation, initiated by the 'planting lairds' of the eighteenth century, has since accelerated with the development of new techniques and machinery and the introduction of new species and varieties; planting has expanded uphill to a maximum of approximately 500 metres and on to wet mineral soils and deep peatlands. Heather moorlands, rotationally burned and traditionally associated with grouse-moors, dominate the drier northern and eastern areas, contrasting with the green braes of Angus and parts of Perthshire. In these southern hills, sheep densities are usually much higher. Above the planting limits, the rough grazings constitute the deer-forests, some of which are maintained exclusively for stalking red deer.

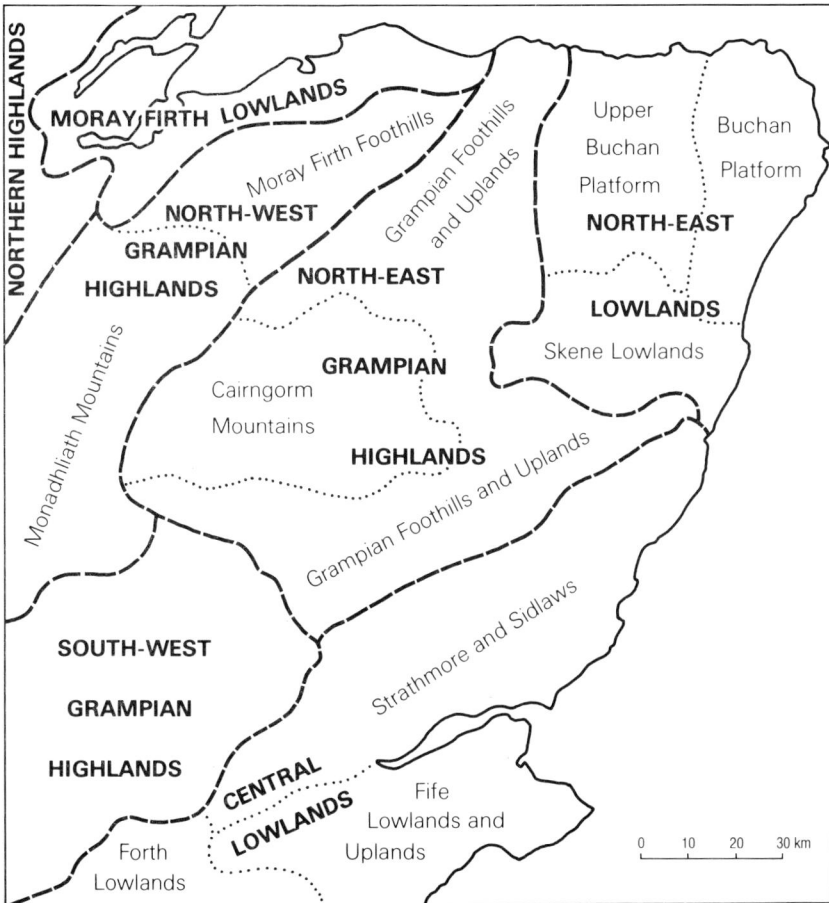

Figure 2. *Physiographic regions.*

PHYSIOGRAPH REGIONS—GEOLOGY, LANDFORMS, PARENT MATERIALS AND SOILS

There are seven main physiographic regions (Fig. 2.):

1 The Moray-Firth Lowlands   4 The North-East Grampian Highlands
2 The North-East Lowlands   5 The North-West Grampian Highlands
3 The Central Lowlands   6 The South-West Grampian Highlands
     7 The Northern Highlands

The distribution of the principal rock types is shown in Fig. 3.

THE MORAY FIRTH LOWLANDS

The Moray Firth Lowlands (Fig. 2) within Eastern Scotland stretch from Nigg in the north to Buckie and lie mainly below 60 metres though the surrounding foothills rise to above 200 metres. Basically, the Black Isle peninsula separates the

3

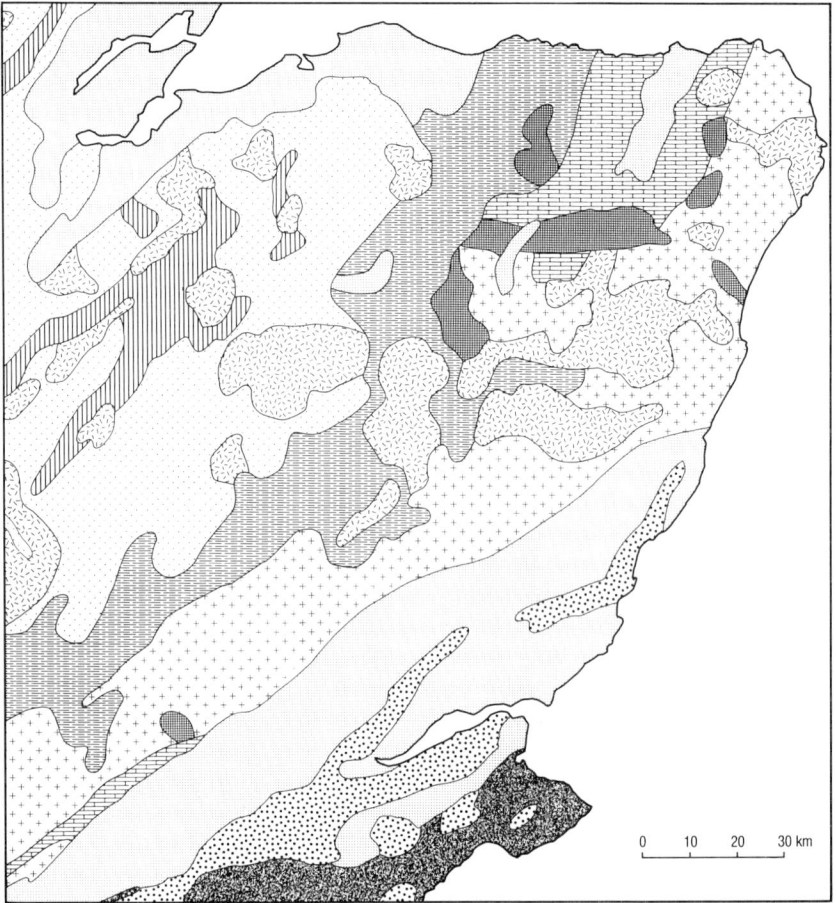

Figure 3. *Geology.*

Cromarty Firth from the Moray Firth. Linking the bases of these firths is a narrow strip containing the lower valleys of the Rivers Conon, Peffrey and Beauly.

Most of the region has been carved from Old Red Sandstone sediments which are bounded by a discontinuous rim of breccias and conglomerates resting unconformably on Moinian rocks, (Fig. 3). In the west, residual conglomerates frequently form inselberg-like hills and provide the reddish brown, shallow, coarse-textured, stony parent material of the Kessock Association. East of the River Spey, however, Middle Old Red Sandstone conglomerates and sandstones form a bright red mixed till with a sandy loam or loam texture, the parent material of the Tynet Association.

Apart from a western belt of Lower Old Red Sandstone rocks stretching northwards from The Aird, the sediments are dominantly red and yellow sandstones of the Middle and Upper Old Red Sandstone divisions. Commonly, the associated drift is a reddish brown, sandy loam or loam till. East of Forres, the till is derived from the Upper Old Red Sandstone formation and forms the parent

4

Sandstones, shales, grits and limestones	Carboniferous	Sedimentary rocks
Sandstones, conglomerates and shales	Old Red Sandstone	
Andesites, basalts and tuffs	Extrusive	Igneous rocks
Granite and allied rocks	Intrusive	
Gabbro and allied rocks		
Quartz-schists, slates, phyllites, graphite schists, epidiorites, hornblende-schists and limestones	Dalradian	Metamorphic rocks
Grits, slates and phyllites		
Quartz-mica-schists, quartzites and grits		
Semi-pelitic and pelitic schists with calcareous and quartzite bands	Moinian	
Granulites, schists and gneisses		

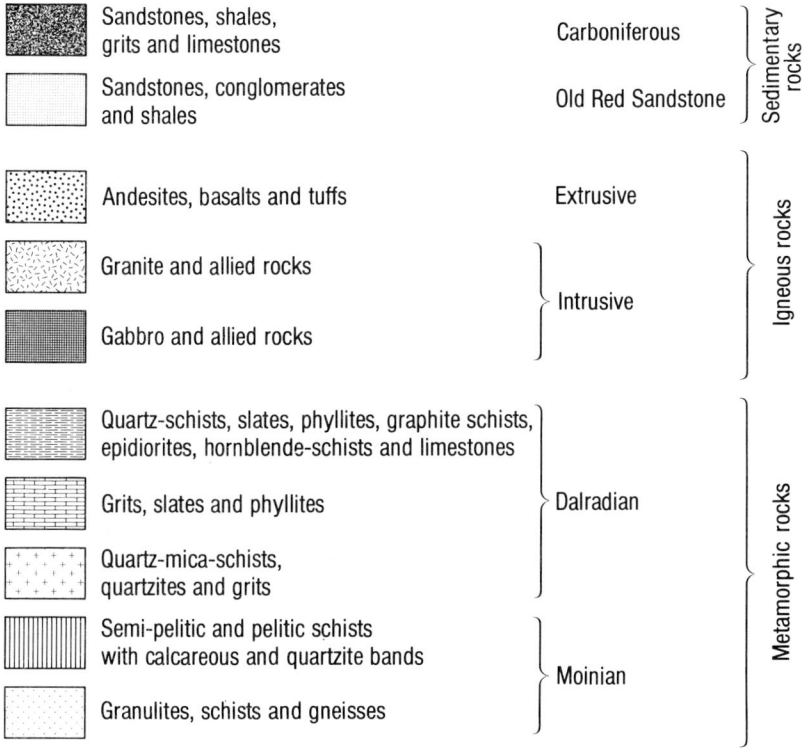

material of the Elgin Association whereas, near Nigg in Easter Ross and around Cromarty, a closely related till, derived mainly from the Middle Old Red Sandstone rocks, forms that of the Cromarty Association. Frequently, these tills are overlain by water-worked or morainic gravelly loamy sands with the combined drifts forming the parent material of the Kindeace Association.

Fringing the western lowlands is the Struie Group of the Lower Old Red Sandstone division which includes drab grey and brown mudstones and siltstones (Armstrong, 1977). Between Strathconon and the northern edge of Eastern Scotland, these sediments, often deeply weathered, form the fine-textured parent material of the Braemore Association. Around Auldearn, a similar but bright red drift, derived from the Middle Old Red Sandstone formation, forms the parent material of the Kinsteary Association. Originally, the western sedimentary rocks, including the marginal conglomerates, were allocated to the Middle Old Red Sandstone division. Recent research, however, indicates that they are better correlated with the Lower Old Red Sandstone rocks, (Armstrong, 1977).

The Black Isle peninsula (Plate 2), rising to 256 metres, consists of red and

Plate 2. *Near Strathpeffer, looking eastwards to the Black Isle in the distance. The cultivated land on the valley floor comprises humus-iron podzols of map unit 420 and noncalcareous gleys of map unit 421 (Nigg Association). The gentle slope on the right, bisected by the shelter-belt, carries brown forest soils of map unit 71 (Braemore Association). Both areas are good arable land (mainly Class 3). The scarp in the middle distance comprises Class 5 and Class 6, with Class 4.1 on the plateau above; the soils are humus-iron podzols of map unit 454 (Mounteagle Association).* Aerofilms

yellow sandstones of the Strath Rory Group of the Middle Old Red Sandstone division (Armstrong, 1977); they form a broad syncline with the axis aligned north-east to south-west. On the spine where the sandstones are close to the surface, the shallow, stony, coarse-textured drift forms the parent material of the Mount Eagle Association. Similar drifts occur on the hills north of Alness and on the western sector of Drummossie Muir.

Along the southern shores from Kessock to Nigg and beyond, the extension of the Great Glen Fault has produced a remarkably straight coastline. Relative dextral and sinistral movements required to match certain structures along the fault were first proposed by Kennedy (1946); the ensuing controversial investigations are summarized by Smith (1977).

Around Rosemarkie and Cromarty, inliers of metamorphic rocks with common veins of red microgranite form high cliffs; the Moinian psammites at Rosemarkie alternate with hornblende-gneiss of Lewisian derivation (Harris, 1977). Collectively, the derived shallow stony drifts form the Ethie Association. Other minor lithologies include tiny outcrops of Jurassic rocks near Ethie and Lossiemouth and Permo-Triassic sandstones near Elgin. The Jurassic-derived soils are too small to accommodate at the 1:250 000 scale and those on materials derived from the Permo-Triassic have been incorporated within the Elgin Association.

Although the chronology of the glaciation is still in question, it is believed that the Moray Firth Lowlands were influenced during the Würm period by at least two ice-sheets. From a centre in the Northern Highlands, ice flowed across the Moray Firth and deflected eastwards ice moving north along the Great Glen; the passage of these ice streams is indicated by the distinctive Inchbae augen-gneiss and Strath Errick granites (Small *et al.*, 1971). Simultaneously, ice moved along Speyside from the Moor of Rannoch to meet the Moray ice near the boundary of the 305-metre peneplain and the foothills; the oscillatory nature of the fronts is shown by the interplay of massive deposits of a brown schist and granite till and a red sandstone till south of Cawdor. The deglaciation, detailed by Synge (1956) and Smith (1977), amongst others, resulted in vast fluvioglacial deposits ranging widely in degree of sorting and texture. Mostly, they are bedded and well-graded sands and gravels of the Boyndie/Corby Associations laid down in a variety of forms including deltas, terraces, pitted outwash plains, eskers and kames. Minor areas of silts and silty clays, probably formed in proglacial lakes, respectively form the parent material of the Craigellachie/Polfaden and Carden Associations. On the southern lowlands, limited areas of light reddish brown, partially stratified, gravelly loamy sands with a mainly sandstone content form the parent material of the Brightmony Association; usually they overlie the red-brown sandstone till below 2 metres. A similarly stratified, pale brown drift with a dominantly metamorphic stone content forms the soils of the Dulsie Association which fringes the foothills.

Apparently confined to the Black Isle are extensive deposits of a coarse-textured moraine derived from sandstone and forming the parent material of the Millbuie Association. Similar drifts, but with a mixed stone content and affinities with the Corby Association, occur throughout the remaining lowlands, mainly below 100 metres.

Most of the low ground below 30 metres has been shaped by eustatic changes in sea level and isostatic uplifts following deglaciation in late-glacial and post-glacial time (Fig. 4). The pulsatory emergence and evolution of these raised beaches have been described by Ogilvie (1923) and Steers (1937). On the higher beaches, the parent materials are mainly sands and gravels, probably of fluvioglacial origin but reworked by marine agencies. They are virtually indistinguishable from the deltaic deposits, for example, around Inverness, Beauly and Muir of Ord, from the associated river terraces or from the pitted outwash deposits which mask much of the lowlands above the highest beaches especially east of Inverness. Collectively, they form the Corby/Boyndie Associations. At the lower beach levels, there is a whole array of spectacular wave-built and wind-built land ranging from the prominent storm-beach shingle west of Speymouth to the dunes at Culbin. Raised beach and estuarine alluviums, commonly poorly drained, are widespread. With textures ranging from sand to silt, these immature soils, together with the shingle, form the undifferentiated raised beach deposits of the Nigg Association; the aeolian sands form the parent materials of the Links Association. In the Spynie basin, silty clays form the soils of the Duffus Association.

The dominant soils of the Moray Firth Lowlands are podzols derived from acid parent materials and characterized by coarse and moderately coarse textures with free or excessive drainage. Most of the region is subject to less than 900 millimetres rainfall, the coastal zones receiving less than 700 millimetres. Below

7

100 metres, the podzols include humus, iron and humus-iron major soil sub-groups though many are now cultivated with the diagnostic upper horizons obliterated; occasionally there are plaggen topsoils up to 50 centimetres thick. Cultivation has been widespread since Neolithic times, especially on the raised beach and fluvioglacial soils.

Humus-iron podzols and peaty podzols usually feature an iron pan above a friable B horizon although in the Boyndie Association the B horizon is occasionally cemented. Induration is widespread in most of the tills and associated drifts, ranging, for example, from a horizon 20 centimetres thick in the Mount Eagle Association to 75 centimetres in the Kindeace Association. A strong, very coarse platy structure is usually developed. Some tills, especially those of the Cromarty, Elgin and Kindeace Associations, often display polygonal cracks in the C horizon surface; probably of periglacial origin, these are about 1 centimetre wide and taper downwards over 1 metre.

Wet mineral soils are confined mainly to estuaries, ex-lagoonal sites and low raised beaches. In addition, there are minor basin peats which occasionally have associated marl bands. Similar marls occur in tiny areas of gleyed lacustrine soils which are too small to identify at the 1:250 000 scale. Apart from the related calcareous gleys of the Duffus Association, most gleys are noncalcareous. Amongst the tills, gleys are of limited occurrence with calcareous gleys found only in the fine-textured soils of the Braemore Association. Humic gleys are usually restricted to flushes and peaty gleys confined normally to the higher elevations.

Because of the common, single-grain structure of the fluvioglacial, raised beach and alluvial soils, poor soil stability is widespread. Compounded by the warm south-westerlies in spring, wind erosion is often a serious hazard to cultivation. The mobile Culbin Sands are a potent reminder of the dangers despite their stabilization by afforestation. Associated also with the unconsolidated nature of the coastal deposits is the constant evolution of the lowest raised beach. Along the southern shores especially, the east–west tides are engaged in a continual cycle of deposition and erosion. Notably, Burghead Bay is being eroded, Findhorn Bay is silting up and the Nairn bar is extending westwards at the expense of the north-eastern tip of Culbin.

THE NORTH-EAST LOWLANDS

The lowlands of the north-east (Fig. 2) project into the North Sea between the Old Red Sandstone lowlands of Moray and Strathmore. It is a complex region, drained by the Dee, Don, Ythan, North and South Ugie, Deveron and Spey rivers, all of which rise in the Grampian Highlands. Despite the smoothing action of erosion across the complex lithology, several major physiographic subregions are clearly defined. They include the Upper Buchan and Buchan Platforms, and the Skene Lowlands. The origin of these platforms and the nature of the erosional agencies are still unresolved. Much of the region has a very gently rolling or undulating topography with some areas of the Buchan Platform being flat and monotonous (Fig. 4). Exceptions are the foothills fringing the Upper Buchan Platform and the Skene Lowlands.

Glaciation appears to have been weak over the greater part of the Lowlands and, according to Synge (1956) and Charlesworth (1956), part of Buchan may not have been glaciated during more recent ice movements. However, field observations by Glentworth (1954, 1963) showed a veneer of drift throughout the region, varying up to 10 metres thick and ranging in texture from a stony, sandy

Plate 3. *Lowlands near Macduff: the scarcity of trees and the exposed gently rolling landscape are typical of much of Buchan. Most soils are cultivated humus-iron podzols of map units 243 (Foudland Association) and 97 (Boyndie Association). This is good arable land, mainly Class 3.1, used for the production of cereals and fat cattle.* SDD Crown copyright.

loam to a clay loam or clay. On hills and local prominences, where it is seldom more than 1 metre and is often less than 0.5 metres thick, it is usually a stony, sandy loam.

The tills which are characteristic of lower slopes have a higher clay content and sometimes a greater variety of stones than the shallow tills or drifts which is usually underlain by shattered or weathered rock. Many areas of drift have been modified by the meltwater of the retreating ice-sheets. Fluvioglacial deposits are generally sands or gravels derived from granitic or metamorphic materials.

Ice-movement across the area emanated from ice-caps in the mountains to the north-west and south-west of the region, but the presence of the Scandinavian ice-sheet off the east coast affected the direction of the movement. The first ice-sheet moved north-west to south-east, depositing in the area of Whitehills in Banffshire the black Jurassic clay from Brora, Sutherland and probably the bed of the Moray Firth. Marine silts and clays of a possible interglacial period occur at Sandend, Banffshire. The second movement was from south-east to the north, and was likely responsible for the transportation of the red Strathmore drift from south of the Highland Boundary Fault to the lowlands of the north-east. Soil survey evidence supports the view that the drift was pushed in from the east by the

9

Scandinavian ice in the North Sea (Glentworth, 1954). This drift is the parent material of the clay soils of the Peterhead Association, the water-sorted soils of the Collieston Association and the lacustrine silts and clays of the Tipperty Association. These three associations occupy nearly 200 square kilometres. The third ice-sheet came from the west and north-west, moving east and south-east along the valleys. It was less extensive than the first or second ice-sheet and did not remove the Strathmore drift from the region.

In the lowlands of the north-east, free-draining soils occupy the flanks and summits of the gently undulating hills; imperfectly drained soils are on the gentle lower slopes; poorly and very poorly drained soils are mainly in depressions where the drift or till cover is thicker and finer in texture. The climatic and geological conditions of the north-east tend to produce acid and podzolic soils. These are of general occurrence in the region whereas brown forest soils are more restricted and are seldom found above 300 metres, except on base-rich parent materials. For example, brown forest soils of the Insch Association exist at altitudes over 300 metres in the Rhynie and Cabrach districts.

The larger areas of brown forest soils in the region belong to the Tarves Association, but considerable areas are found also within the Insch and Tipperty Associations. Gleys are especially extensive within the Tarves Association on the Buchan Platform, whereas the main areas of podzols occur within the Foudland Association on the Upper Buchan Platform and the Countesswells Association in the Skene Lowlands. Because of their gently rolling topography, favourable climate and soils, most of the lowlands of the north-east have been extensively cultivated for centuries.

The Buchan Platform consists of a peneplain tilted to the east and largely covered by glacial drift. It stretches from Aberdeen to the Moray Firth with re-entrants into the Upper Buchan Platform along the North Ugie, South Ugie and Ythan valleys. The topography is gently undulating and the land intensively farmed. Apart from minor wooded areas, for example, near Tarves, Old Deer and Mormond Hill, it is singularly devoid of forest plantations.

The underlying rocks are mainly folded, resistant metamorphics of the Dalradian Assemblage aligned north-north-east to south-south-west (Fig. 3). They include quartz-mica-schists, grits, quartzites, slates and phyllites with the prominent hills formed from the more resistant rocks. Mormond Hill, the most prominent hill on the Buchan Platform, is formed from quartzite and is 240 metres high.

The quartz-mica-schists and coarser, interbedded, quartzose bands form the parent material of the Strichen Association, whereas the slates and phyllites are the principal parent rocks of the Foudland Association. Quartzites are parent rocks for the Durnhill Association. At Maud, Tarves, Arnage and Belhelvie there are small gabbro intrusions related to the Caledonian Orogeny, mainly contaminated with small xenoliths of the country rocks (Read and MacGregor, 1948). On the tops of hills, these basic rock types yield a parent material approximating that of the Insch Association. The main areas underlain by such rocks, however, have a cover of mixed drift and the derived soils belong to the extensive Tarves Association. Such a mixed drift is also widespread between Newmachar and Tarves where it shows the influence of the underlying gneiss; and from Udny to Longside where the underlying granite, quartz-schist and gneiss formations have all contributed to the parent material. Although derived from many different rock types, the till of the Tarves Association has major contributions from both acid and basic rocks with the influence of the acid rocks invariably the greater.

10

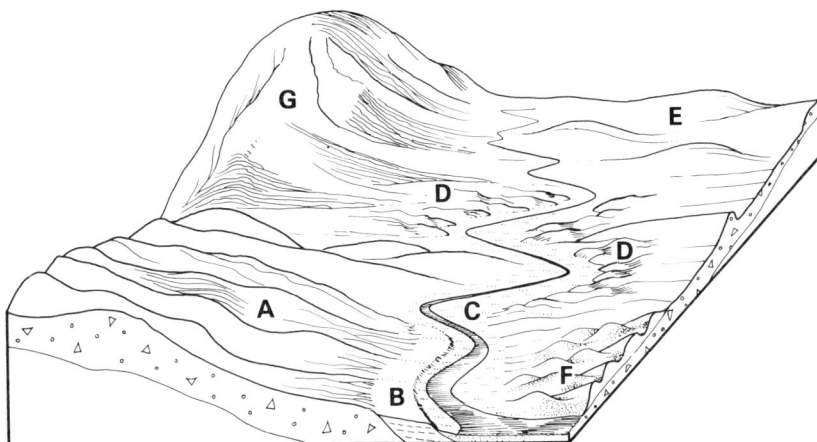

A Till slopes in lowlands and foothills
B Raised beach
C Alluvial terrace
D Fluvioglacial terrace
E Till plain with isolated mounds
F Dune complex
G Isolated hill with till cover on lower slopes

Figure 4. *Landforms of the North-East and Moray Firth Lowlands*

Two major intrusions of granite crop out within the Buchan Platform, the coarse, flesh-coloured, silica-rich Peterhead granite and the fairly coarse, light-grey granite of the New Pitsligo area. These granites and granitic gneisses form the parent material of the Countesswells Association.

The Upper Buchan Platform extends along the western edge of the Buchan Platform. It is underlain by Dalradian rocks and consists of a broadly rolling terrain with hills which tend to be aligned in a north-west to south-easterly direction (Fig. 3). The smooth slopes are well drained and covered by a thin drift, the whole platform being extensively cultivated (Plate 3).

On the western edge of the Upper Buchan Platform, between Banff and Huntly, there is a narrow band of the Keith Division of the Dalradian. As with the Buchan Platform, these rocks form the parent material of the Strichen Association. Between Gardenstown and Fyvie, the underlying rock is the Gamrie–Turriff outlier of the Middle Old Red Sandstone. Within this succession is a dominantly coarse conglomerate, with pebbles and boulders of schist and quartzite up to 40 centimetres but also including marls and associated sandstones. This forms a stony sandy loam, loam or sandy clay loam till, the parent material of the Hatton Association. The red sandstones are the main parent rocks of the sandy loam, loam or sandy clay loam till of the Cuminestown Association. A basal conglomerate consists of argillaceous schists, mainly slates and grits from the Banff Division of the Dalradian on which it rests. This mixture provides the parent material on which soils of the Ordley Association are developed.

Two relatively small areas of almost pure quartzite gravels around Fyvie may be Pliocene gravels of the Tertiary Period. The rest of the Upper Buchan

11

Platform is underlain by the eastern or Banff Division; it comprises Dalradian schists of low-grade metamorphism from which is formed the fine sandy loam till and drift of the Foudland Association.

The Insch valley has resulted almost entirely from the weathering of a plutonic basic igneous mass although the western end is formed within the Middle Old Red Sandstone outlier which extends southwards past Rhynie to Kildrummy. The erosion of these readily weathered rocks has produced the basin, which is 13 kilometres wide and 50 kilometres long. It has an altitude range from 60 to 250 metres and extends from the Bogie valley in the west to Oldmeldrum in the east. The basic igneous rock forms the sandy loam or sandy clay loam drift, between 1 and 10 metres thick, on which soils of the Insch Association are developed. These soils are inherently fertile and have been intensively cultivated for many centuries. The area has a high reputation for the quality of its crops and stock; an abundance of summer grazing is available on the hills and the arable land at a lower elevation provides winter-keep for the stock.

The Skene Lowlands extend from Aberdeen westwards to the Grampian Foothills. From Tarland to Aberdeen, this area is underlain almost entirely by various granites (Fig. 3). The Hill of Fare, for example, is formed from a pink or flesh-coloured acid granite of uniform texture, with small outcrops of a similar granite at Tarland, Alford, Cairn William and Coldstone. The remainder of the granite mass, referred to as the Skene Complex, is a grey granite with a variable composition. A greyish pink granite, intermediate in colour, has been quarried at Tyrebagger and Clinterty, whereas at Corrennie both the pink and grey granites occur in the same quarry. To the north and south of Aberdeen, and also around Inverurie, the underlying rock is more gneissose.

The granite and granitic gneiss rocks form the parent material of the Countesswells Association. On convex slopes, the till or drift is a coarse sandy loam with many stones and boulders, whereas on concave situations, it is usually a loam or sandy clay loam till with comparatively fewer stones and boulders.

Because of the rolling topography with long smooth slopes, a moderately favourable climate and predominance of freely and imperfectly drained soils, most of the Skene Lowlands has been extensively cultivated for a long time. Many of the hill summits are now planted with trees. The reclamation achievements of the north-east farmer may be seen from the number of 'consumption' dykes of granite boulders in the areas of the Countesswells Association in the Skene district. Some of these stone clearance walls may date back to the fourteenth century, although the peak period of development would have been from the eighteenth to the nineteenth century.

There are extensive sand and gravel deposits throughout the Skene Lowlands and both the Dee and Don rivers have large, recent alluvial flood plains. The channels of both rivers often have high flanking fluvioglacial sand and gravel terraces. These fluvioglacial deposits form the parent material for soils of the Corby/Boyndie/Dinnet Associations.

The Alford and Tarland basins are comparatively large depressions within the Skene Lowlands and are almost surrounded by the Grampian Foothills. They are isolated areas of good arable land. The Alford basin covers an area of approximately 50 square kilometres. From the River Don, the slopes very gradually steepen with gentle undulations until they reach the strong or steep slopes of the surrounding hills. Outwith the flood plain, the basin is covered by a thick drift of the Tarves Association. On the somewhat flatter, eastern side of the basin, the soils tend to have poor natural drainage.

The Tarland basin is crescentic in shape, the western end opening into the valley of the Dee. The eastern end is drained by the Tarland Burn which has wide stretches of associated alluvium. Sheltered by the high hills to the north and east, for example, Pressendye (600 metres) and Craiglich (450 metres), the basin is bounded by lower hills on the south and west. The slopes tend to rise more steeply than those of the Alford basin, though the drift mantle covering the slopes contains also a high proportion of basic igneous material, forming the Tarves Association. The soils are of a good agricultural quality and their capability is enhanced by the south-facing aspect which permits cultivation on the higher slopes.

THE CENTRAL LOWLANDS

The Central Lowlands (Fig. 2) are situated to the south-east of the Highland Boundary Fault which stretches within Eastern Scotland (Sheet 5) from Stonehaven to Loch Lomond. The southern boundary is the lower edge of the mapped area extending from Loch Lomond to Kirkcaldy, on the southern shore of Fife. This broad, undulating lowland occupies the northern section of the much larger Midland Valley of Scotland (Plate 4). The area is structurally part of a rift valley let down between parallel faults and the region is limited in altitude in contrast with the great mass of the Grampian Mountains to the north and the Southern Uplands to the south. Whereas the greater part of the region lies below 175 metres, many areas exceed 300 metres and the highest parts rise to 600 metres. In this respect, the regional relief has been determined basically by differential erosion of the rock types with the igneous rocks, which are the most resistant, forming hill masses, for example, the Ochil and Sidlaw Hills, and the softer, sedimentary rocks underlying broad plains, for example, Strathmore, the Howe of the Mearns and Fife (Fig. 3). The landscape has been modified, however, by the effects of glaciation and this has resulted in the formation of many different soil parent materials Laing (1976).

Initiated during the Caledonian Orogeny, the Central Lowlands formed a basin of deposition for Old Red Sandstone sediments, subdivided into Upper and Lower Old Red Sandstone strata. The latter comprise the thicker and more widespread formation. Structurally, the Lower Old Red Sandstone sediments form a syncline, which reaches its maximum width between Methven and Perth, a region known locally as Strathmore. Smaller basins occur within the Howe of the Mearns, at Strathearn and Strathallan and beneath the Carse of Stirling. The sediments were deposited under semi-arid, fluviatile and lacustrine conditions and gave rise to a succession of dull-red sandstones, which are coarse grained and massively bedded, followed by coarse conglomerates, flagstones and occasional limestones, marls and mudstones.

Nearest the Highland Boundary Fault, the outcrops of basal conglomerates have tended to resist erosion and are associated with hilly ground, for example, Strathfinella Hill near Stonehaven, the Hill of Alyth and high ground between Crieff and Callander. The softer marls and mudstones are well developed within small basins in the vicinity of Edzell and Laurencekirk, and in small areas near Perth, Gartmore and Methven. During the deposition of all these sediments, there was considerable volcanic activity which resulted in widespread igneous intrusions. Where the harder sandstones, conglomerates and lava strata protrude, a series of long, low, parallel ridges has developed. The Garvock and Aberlemno Hills rise to 300 metres and Rossie Muir, which is underlain by lava,

Plate 4. *Kippen Muir viewed from the foothills south of the Forth valley with Ben Vorlich and Stuc a'Chroin in the background. In the foreground, the noncalcareous gleys and brown forest soils of map unit 467 (Sorn Association, Class 4) contrast sharply with the peaty gleys, peat and peaty podzols of Kippen Muir in the middle distance, map unit 344 (Kippen Association, Class 5). On the gentler slopes across the Muir, brown forest soils of map unit 274 (Gourdie Association, Class 5) are dominant. The humus-iron podzols and peaty podzols of map unit 493 (Stonehaven Association, Class 6) on the higher steeper slopes are frequently afforested.* Institute of Geological Sciences photograph published by permission of the Director; NERC copyright.

forms the summit of a prominent, broad area of rising ground separating the Montrose Basin from Lunan Bay. Higher ground is formed by the more distinctive, andesitic and basaltic lavas and tuffs of the Sidlaw (408 metres) and Ochil (788 metres) Hills. These hills are cut by the Rivers Tay and Earn, the northern section continuing into Perthshire and Angus as the Sidlaw Hills and the southern section, or Ochil Hills, stretching from the northern fringe of the Fife peninsula to Stirling. Both hill masses are craggy, often with stepped rock outcrops.

Sandstones, conglomerates and mudstones of Upper Old Red Sandstone age outcrop locally across the Central Lowlands but are comparatively unimportant. They underlie, however, the Howe of Fife, the Kinross Plain and Stratheden. To the north of the Ochil Hills, the same strata underlie part of the Carse of Gowrie, along the Firth of Tay and small inliers appear on the Forfar–Kincardine coast in the neighbourhood of Arbroath and St Cyrus. Contemporaneous igneous rocks are practically unknown within the sedimentary rocks of this period.

Upper Old Red Sandstone strata pass conformably upwards into the Carboniferous strata, the rocks of which form a broad, compound syncline and underlie the lowlands of Stirling, Clackmannan and south-west Fife. These Carboniferous

14

sediments include mainly sandstones and shales with some coals and occasional limestones and calciferous sandstones. Throughout most of the period during which these sediments were being deposited, and extending into the Permian time, considerable volcanic activity took place with the resultant formation of intrusive sheets or bosses, sills, dykes and volcanic necks. To the south of the middle Forth valley, the Gargunnock Hills, which consist of a vast sequence of basaltic lava flows, form the northern tip of the Campsie Fells plateau. The separate lava flows form dip and scarp features and terraced hills which rise steeply in a series of cliffs from the Forth valley. Other basaltic masses, typified by the Saline, Benarty and Cleish Hills, occur in Fife where they form separated groups of hills and detached knolls. To the west of Loch Leven, the Lomond Hills form another significant hill mass. Roughly circular in shape, with steep scarp slopes more than 150 metres high, the mass consists of sandstones, mudstones and limestones, capped by a massive and resistant sheet of dolerite. The peaks of West (522 metres) and East (488 metres) Lomond are the infilled vents of volcanoes which penetrated the mass. It is likely that the Central Lowlands were significantly developed by the mid-Tertiary times and that the only subsequent major influence has been the pronounced effect of the Pleistocene Period on the local landscape.

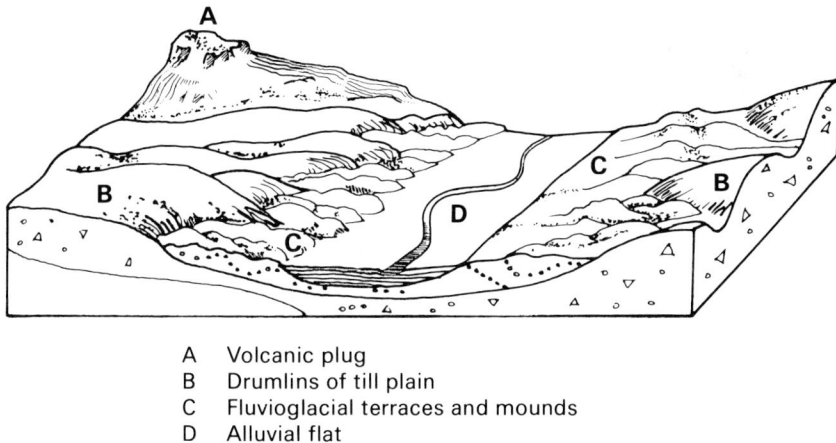

A   Volcanic plug
B   Drumlins of till plain
C   Fluvioglacial terraces and mounds
D   Alluvial flat

Figure 5. *Landforms of the eastern Central Lowlands.*

The glacial history has significant importance because the form, composition and distribution of the soil parent materials and much of the localized topography resulted from that period (Fig. 5). Three distinct phases of the glaciation have been recognized. During the first phase, ice moved south-eastwards from the Highlands into Strathmore but was deflected to the south-south-west by the pressure of the Scandinavian ice-sheet. The resultant glacial till was derived, in part, from the schistose rocks of the Highlands and in part from the underlying sandstones and lavas of the Lowlands, forming the mixed parent material of the Gourdie Association. A similar material, but devoid of lavas, forms the Callander Association. It has been deposited in the west within the Teith valley, across Lennieston Moor and in the upper Forth valley, as hummocky moraines and as a veneer on long, gentle slopes.

The second major ice advance spread in a north-east direction across the Central Lowlands, traversing the Lower Old Red Sandstone rocks and depositing the reddish brown till of the Balrownie Association which covers most of the lowland between the Highland Boundary Fault and the Ochil–Sidlaw hill mass. A similar parent material derived from flaggy sandstones, mudstones and siltstones occurs on the lower slopes of the hills north of Doune, where the topography is generally drumlinoid below 180 metres (Fig. 6). It occurs also on the northern edge of the drumlin field, which extends from Kippen to beyond Gargunnock. Where the ice crossed small areas of marls and fine-grained sandstones, a bright-red till, the Laurencekirk Association, was formed. Till derived from Upper Old Red Sandstone rocks, the parent material of the Kippen Association, is widespread in the drumlin fields south of Kippen, in the northern section of the Kinross plain extending from Loch Leven to the Howe of Fife and in a narrow band of land bordering the Carse of Earn. In these locations, a mixed till, derived from such rocks and sandstones and cementstones of Calciferous Sandstone age, the parent material of the Sorn Association, also occurs. The Carboniferous lowlands are covered by a variety of tills which reflects the complex sequence of the underlying strata. The rocks of the Carboniferous Limestone Series in Stirling and Fife produce a greyish till; till overlying the Coal Measures strata is a purplish red-grey-brown whereas till derived from the Millstone Grit Series is purplish brown-grey. All these tills form the Rowanhill Association and are moderately fine textured, mainly sandy clay loam and clay loam. Because they are developed *in situ* or from locally derived material, the parent materials of the igneous hill areas are relatively uniform. The intermediate andesitic and basaltic lavas and tuffs give rise to distinct forms of till and drift, the parent material of the Sourhope Association, which has been separated from that derived from the basaltic lavas and basic intrusive rocks. These latter rocks form soils of the Darleith Association. The lower slopes of the Sidlaw and Ochil Hills are covered by a till derived from mixed lava and Old Red Sandstone sediments, the Mountboy Association. Similarly, a mixed till of Carboniferous sediments and igneous material, the parent material of the Hindsward Association, occurs in south-east Fife.

Broad low-slope drumlins generally, with occasional steeper variants; occasional peat mosses in larger hollows.

Figure 6. *Landforms of the western Central Lowlands.*

The final phase of the glaciation consisted of temporary advances of the Highland ice and the subsequent retreat of glaciers along the Tay and Forth valleys. According to some authors (Simpson, 1933; Sissons, 1963), the Perth Readvance moved into Strathallan and the Forth valley and into the Earn and Almond valleys as far as the neighbourhood of Perth, creating an outwash plain of sands and gravels which in one section overlie rhythmically-bedded marine sediments. Yet according to others, including Paterson (1974), the same phenomena are interpreted as the formation of an advancing delta; they do not regard extensive fluvioglacial landforms in western Strathmore as providing support for the major readvance hypotheses, a view now accepted by Sissons (1976). The Loch Lomond Readvance, similar in character, has left its mark in the lowlands of the south-west around the Lake of Menteith. Besides the deposition of mixed tills, for example, those of the Gourdie and Callander Associations, the main consequence of the readvances was the deposition of vast amounts of fluvioglacial material, as hummocks, eskers, etc. Moraines, found along the scarp face of the Gargunnock Hills and consisting almost entirely of Carboniferous lava fragments mixed with subrounded Highland erratics, form the parent materials of the Kirktonmoor Association. In the Teith and upper Tay valleys, fluvioglacial sands and gravels of the Doune Association are derived mainly from schist and Old Red Sandstone rocks whereas in the lower Tay and Isla valleys fluvioglacial sands and gravels are composed of rocks of Highland origin; the latter form the parent material of the Corby Association. Moundy fluvioglacial deposits of Old Red Sandstone sediments and lavas, together with some Dalradian rocks, form the Gleneagles Association which is extensive in Strathallan. In north-east Fife, however, the parent material is mainly of Old Red Sandstone derivation, the Auchenblae Association. Sands and gravels composed of Carboniferous sediments, Old Red Sandstone lavas and some Dalradian erratics occur north of Stirling and form the Darvel Association.

Because these different phases of glaciation have been accompanied by oscillations in relative land and sea levels, one principal outcome has been the deposition of raised beach materials. Of the late- and post-glacial raised beaches, the lower raised beach is composed of sand whereas estuarine silts and clays of the Stirling Association form the Carses of Stirling, Gowrie and Earn. The higher raised beaches consist usually of mixed sands and gravels derived from Old Red Sandstone rocks, the parent material of the Panbride Association, or from Carboniferous sediments, the parent material of the Dreghorn Association. Silty fine sand and gravels form the Carpow Association and occur within the Eden valley west of St Andrews, on the southern bank of the River Forth around Carbrook and Bannockburn, and in the west near Arnprior.

Recent deposits include both windblown sands of the Links Association and shelly sands of the Fraserburgh Association. Basin peats are of restricted occurrence, important examples being Flanders and Ochtertyre mosses in the west and Methven moss in the east.

Within the Midland Valley, the climate regions are mainly warm moderately dry lowlands, warm wet lowlands and fairly warm moderately dry lowlands and foothills. The primary effect of the climate upon the parent materials has been the production of moderately leached soils. Brown forest soils, brown forest soils with gleying, and to a lesser extent, humus-iron podzols predominate. In general, these genetic soil subgroups are less easily differentiated in the Central Lowlands because of the history of extensive cultivation there.

Brown forest soils are widespread throughout the area on till deposits and occur in all the common soil associations, for example, Balrownie, Stonehaven, Mountboy and Forfar Associations; they are widespread also in the Darleith and Sourhope Associations which occur mainly on the higher ground and are associated with volcanic rocks. Brown forest soils with gleying are also extensive and are found intermingled with the brown forest soils. The medium and moderately fine textures, usually loam and sandy clay loam, of the above associations tend to impede internal soil drainage on low-angle slopes. The subgroup,

17

however, is most common in Fife where the slightly finer-textured soils, mainly clay loams, of the Rowanhill/Giffnock/Winton Associations are widespread.

Humus-iron podzols are not extensive and many are considered to be of recent origin, the soils having degraded from brown forest soils. The parent materials are generally coarser textured than those of the brown forest soils, for example, the fluvioglacial deposits in Strathmore and the Howe of the Mearns (Corby/Boyndie and Auchenblae Associations), in Strathallan (Gleneagles Association) and in Stratheden (Eckford Association). Although these podzols occur also in the lowlands on tills with water-modified upper horizons, for example, the Forfar Association, they are more commonly associated with the upper slopes of the adjacent foothills, for example, the Gourdie/Callander/ Strathfinella and Sourhope Associations.

Peaty podzols are confined mainly to the higher and wetter areas, above 300 metres and with an average annual rainfall of over 1000 millimetres. They are common in the Ochil Hills (Sourhope and Balrownie Associations) and in the Gargunnock Hills (Darleith Association) where they occur in an altitudinal sequence of humus-iron podzols, peaty podzols and blanket peat. Often the map unit comprises the whole sequence. The profile on these hill sides, which frequently support a *Nardus* grassland, usually has an iron pan above a friable B horizon. Peaty podzols occur also on the steep-sided drumlins (Kippen Association) at the northern edge of the Gargunnock Hills; minor areas of peaty podzols (Stonehaven and Gourdie/Callander Associations) are located also on the steeper slopes of some of the Menteith Hills.

Of the gleys, the noncalcareous gley is the most common major soil subgroup. Within the lowlands, from the extreme south-west to Brechin in the north-east, it is mainly associated with the sandstone-derived till of the Balrownie Association. Two other moderate areas of noncalcareous gleys exist; in the Ochil Hills, they belong to the Sourhope Association whereas to the south, on the west Fife plateau, they are associated with the Carboniferous till of the Giffnock Association. Minor areas include those immediately north of the Gargunnock Hills and belonging to the Sorn and Kippen Associations. All these soils are formed in moderately fine-textured tills, sandy clay loams and clay loams, and occupy depressions. Partly because of the relatively low rainfall and high temperatures coupled with intensive drainage, poorly drained soils do not form a dominant soil type within the eastern Midland Valley; imperfectly drained brown forest soils are much more widespread.

Peaty gleys are smaller in extent than the noncalcareous gleys and are found mainly west of Stirling in the foothills where the rainfall exceeds 1200 millimetres. On the lower Braes of Doune, there is a large complex area consisting of peaty gleys (Balrownie Association) and small basin peats. The only other significant area of peaty gleys (Kippen Association) exists to the south-west of Kippen.

Amongst the less extensive major soil subgroups, brown rankers are confined to the andesitic lava summits (Sourhope Association) on the Ochil Hills and to the stepped, basaltic lavas (Darleith Association) on the northern slopes of the Gargunnock Hills. Regosols are restricted to windblown coastal sands (Links Association), often stabilized. They are locally extensive around bays and at the mouths of estuaries. Calcareous regosols formed in shelly sands (Fraserburgh Association) are significant only in Largo Bay, Fife.

Basin peats are confined largely to the Fife peninsula and to the upper Forth valley. In the latter area, they form remnants of a much more widespread peat mantle which developed on the poorly drained carse clays but was almost completely removed during reclamation in the late eighteenth century. The lower limit of blanket peat is virtually coincident with the 1600 millimetres isohyet; such peat is a feature of the summits of the Ochil Hills and Gargunnock Hills and of the upper slopes of the Braes of Doune.

THE NORTH-EAST GRAMPIAN HIGHLANDS

This upland region is bounded by the Highland Boundary Fault in the south-east, by the River Spey in the north-west, and by Glen Garry and Strath Tay in the south-west (Fig. 2). To the north-east, the region grades into the upper Buchan Platform and the Skene Lowlands, the rather irregular and indistinct boundary approximating to the 300-metre contour.

The highest ground is in the west, formed by the Cairngorm Mountains at 1300 metres and the band of high summits from the Gaick plateau through to Lochnagar. To the north, east and south, hill top elevations fall away to between 400 and 600 metres, most of the region being part of a peneplain with distinct planation surfaces.

The derivation of these platforms from a former Highland peneplain created from 'the worn down remnants of the ancient mountains of the "Grampian Caledonides" together with their mantle of later strata' (Johnstone, 1966) has been examined at length by various authors; they include Geikie (1901), Peach and Horne (1930), Bremner (1915, 1919), Linton (1951), George (1965), Sugden (1968) and Sissons (1976). Peach and Horne, for example, drew attention to the planation surfaces at different levels and postulated an original, easterly tilted surface upon which the consequent rivers flowed. Linton suggested that the maximum elevation was in the Ben Nevis–Cairngorm area and that its arched surface led to the present-day summits declining towards the east and south-east. George, however, proposed a pulsatory uplift later than the period of Tertiary igneous activity. These various hypotheses have been summarized by Sissons. More than a third of the area lies above 600 metres and only a few river valleys lie below 300 metres.

Hills in the region, in contrast with hills farther to the south-west, are mostly rounded with few rock outcrops. Slopes tend to be smooth and convex, rising steeply from valley floors to broad, gently sloping hill summits (Fig. 7).

The main river, the Dee, has a large central catchment area and flows east-wards, as does its smaller, meandering neighbour, the River Don. Tributaries of the Spey and Deveron provide drainage to the north, whereas rivers of the Angus glens, together with several tributaries of the River Tay, flow south and south-east, draining the southern portion of the region.

During the Pleistocene glaciation, the region was inundated by ice moving out radially from Rannoch Moor, the main Highland centre of accumulation. Local ice-caps on the Cairngorm Mountains and other high plateaux and the arrange-ment of pre-existing valleys had, however, a considerable influence on the direction of ice-movement.

Apart from removal of the preglacial solum, there are spectacular, localized features demonstrating the powers of glacial erosion, for example, the corries of the Cairngorm Mountains, the glacially deepened trough of Glen Clova, the glacial breaches of the Lairig Ghru and upper Glen Derry, and the meltwater gorges of Ailnack and the upper Avon (Plate 5).

Deposits of glacial drift are widespread, being normally thickest in the valleys and thinning out upslope. Accumulations of till are most common among the lower hills of the region, notably in Perthshire and Angus. In the more

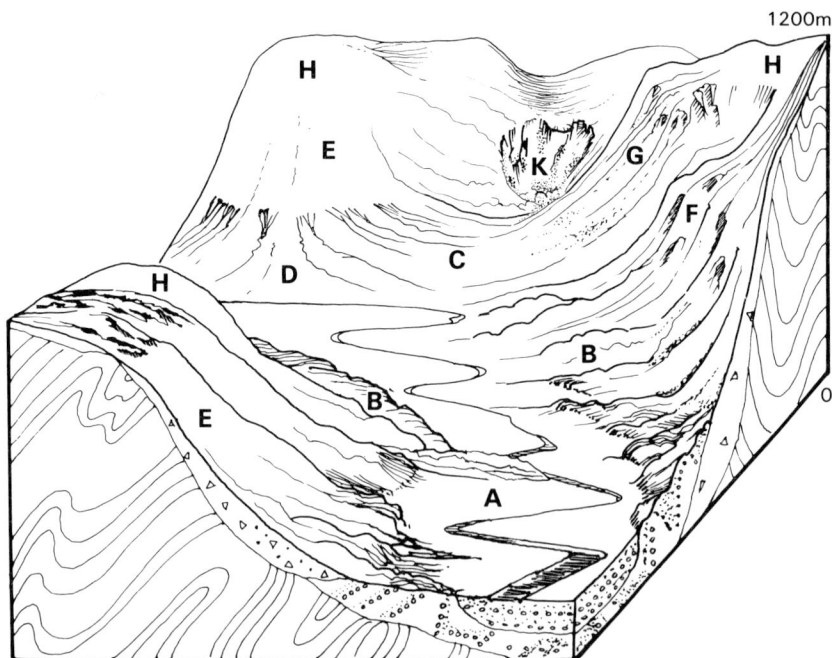

A Alluvial flats and terraces
B Outwash mounds, ridges and terraces
C Moraines in mountain valley sites
D Till slopes in foothill and lowland sites
E Hill sides with steep and very steep slopes; non-rocky
F Hill sides with steep and very steep slopes; moderately rocky
G Hill sides with steep and very steep slopes; very rocky
H Rounded hills and plateau summits
K Corrie

Figure 7. *Landforms of the North-East Grampian Highlands.*

mountainous areas, till is found occasionally as valley infill and in embayments among hill slopes. Generally, however, the hills are covered by thin, stony, locally derived, mostly coarse-textured drift which overlies shattered bed-rock at depths of less than 2 metres. This type of soil parent material is probably the most abundant in the region. Widespread but sporadically distributed pockets of weathered rock, with or without a drift cover, occur among all lithologies. Moundy morainic deposits, though widely distributed, are not as common as in the valleys of the Western Highlands. Good examples occur in Glen Callater and in Caenlochan south of Braemar.

Deposits of sand and gravel occupy many major river valleys. Fluvioglacial terraces, frequently associated with recent alluvium, form ribbons of cultivated land in the Dee valley and in the Angus and Perthshire glens. Kame and kettle-hole topography, more localized, occurs, for example, on Deeside near Dinnet and at Ballochbuie, and on Speyside at Abernethy and Rothiemurchus. Peri-glacial conditions probably persisted during deglaciation at the last major ice-

Plate 5. *Loch Avon, looking westwards across the Cairngorm plateau. The corries of Braeriach and Cairn Toul are in the upper left background. Surrounding Loch Avon (725 metres) the discontinuous rim of granite crags and the mainly stabilized, scree slopes are some 300 metres high. On the plateau, especially between the loch and Coire an t'Sneachda (in the upper right), are prominent boulder lobes. Above the western end of the loch lie the snow-beds of the Feith Buidhe dominated by* Nardus stricta, Carex bigelowii *and* Deschampsia flexuosa. *Elsewhere the vegetation is sparse and characterized by* Juncus trifidus *and* Racomitrium lanuginosum. *The soils are dominantly lithosols and rankers of map unit 137 (Countesswells Association, Class 7). Where the mantle of granite detritus is sufficiently thick, alpine soils are developed. Subalpine soils are formed in the colluvium and moraine around the shores. The climate is extremely severe and the plateau environment is periglacial.* Aerofilms.

sheet and during the Loch Lomond Readvance. In the latter period, the Western Highlands experienced widespread active glaciation, but in the North-East Grampian Highlands, probably only a few local glaciers were formed, mainly in the corries of the Cairngorm Mountains and in the Lochnagar and Glen Clova areas. This left most of the region exposed to the action of freeze-thaw and frost, causing disintegration of rock exposures and the formation of boulder lobes and stripes. Solifluction, together with these other periglacial processes, helped to produce the smooth and rounded landscape.

21

Geologically, the region is underlain by pre-Cambrian schists into which have been intruded a number of granite plutons (Fig. 3). In the west, granulites and quartz-mica-schists of the Moine Assemblage form the Gaick plateau and the Cromdale Hills and the parent rocks of soils of the Arkaig Association. The diverse Lower Dalradian suite occupies a belt up to 20 kilometres wide, from Cullen in the north to Loch Lomond in the south-west. It consists of sinuous bands of mica-schists, quartzites, black and slaty schists, hornblende-schists and limestones, these being respectively the parent rocks for soils of the Strichen, Durnhill, Foudland, Tarves and Deecastle Associations. Collectively, they form a complex zone of soil parent materials north-east of the Cairngorms and in the Glenshee area. The erosion-resistant rocks, quartzite especially, form hill ridges aligned roughly south-west to north-east, parallel to the regional strike. Common features of the quartzite hills, for example, Corryhabbie, are boulder garlands and screes on steep slopes due to a tendency of the rock to shatter into sharp-edged, angular blocks. In contrast, mica-schist and black schist fragment into small, platy stones and these rocks give rise to smooth, largely rock- and boulder-free hills, such as the Ladder and Correen Hills. The less varied Upper Dalradian rocks, quartz-mica-schists, and grits, occupy the Highland edge and the Angus glens where the main valleys are perpendicular to the regional strike. Soils developed in materials derived from those rocks are included in the Strichen Association.

Granite intrusions, with their resistance to erosion, have formed the Cairngorm massif and several prominent hill masses in the region including Lochnagar, Ben Rinnes, the Mounth area and the outlying Bennachie and Hill of Fare. A feature of these granite landscapes is the abundance of subrounded granite boulders. Soils developed from granite material belong to the Countesswells Association. Less common rocks include the basic igneous rocks, gabbros and norites, of the Morven area which give rise to soils of the Insch Association; in the Cabrach and Strathdon areas, serpentinite occurs, the parent rock of the Leslie Association.

The distribution and pattern of major soil subgroups are influenced by several factors. There is the geographical specificity of soil parent material and its inherent chemistry which, in the hills of the North-East Grampian Highlands, is strongly related to the underlying rock. In this upland region, another powerful factor in soil development is the local climatic variation related to altitude. With increase in height, the resultant higher rainfall, lower temperature and greater exposure have pronounced effects on biotic, physical and chemical activities in the soil. Superimposed on these factors is the overall rainfall trend, the increase westwards encouraging the development of wet and peaty soils.

Podzolic soils, mostly freely drained, are dominant in this region because of the acid and coarse-textured nature of the parent materials and the cool, moderately dry climate. Humus-iron podzols, locally cultivated, typify the foothills and hill-slopes of the drier areas. As altitudes rise westwards, they become restricted to valley floors and the adjoining steep, lower slopes, and are rarely found above 500 metres. Humus-iron podzols are also widespread on the less acid, soil parent materials, such as slaty schist, where their altitudinal limit is correspondingly higher.

Where soil or climatic conditions result in the build-up of a peaty layer, peaty podzols are normally found. Among the lower hills, they frequently occupy the higher, often gentler slopes above the humus-iron podzol zone, but further west they become the dominant soils below 600 metres on all but the steepest slopes and can rise to 850 metres. On the base-deficient soil parent materials such as quartzite and granite drifts, peaty podzols are predominant throughout the boreal zone. Peaty podzols which are gleyed above an iron pan occur more frequently in the wetter west, forming readily in the acid parent materials.

The zone of optimum blanket peat development lies between 550 and 750 metres. In the east, peat covers many of the gently sloping hill tops above the upper limit of peaty podzols. Farther west, it is frequently and intimately associated with peaty podzols and with higher, level sites but rarely occurs on slopes greater than about 15 degrees. Peat associated with alpine and subalpine soils is restricted mainly to hollows, cols, shoulders and other sheltered sites. At these higher altitudes, the peat is invariably hagged.

Subalpine soils, mostly freely drained podzolic soils, occur above 500 metres where the vegetation is wind-cropped due to exposure. They are found often on steep slopes with prolonged snow cover and tend to be more common on north-facing slopes. Subalpine soils are less extensive in the North-East Grampian Highlands than in upland regions farther west because of the smaller amplitude of the terrain and the relatively short upper hill slopes.

Alpine soils are extensive in the region, the Cairngorm plateau forming the largest single area in Scotland. Elsewhere, highly exposed hill summits with alpine soils are found as low as 500 metres. The higher sites are often only partly vegetated, and where summits are swept clear of snow, the soils are subjected to freeze-thaw processes for several months. Alpine soils are overwhelmingly freely drained and podzolic but a few high-level, receiving sites support gleys.

The distribution of other major soil subgroups in the region is less related to altitudinal zones. Brown forest soils, including brown podzolic soils, (Ragg *et al.*, 1978) occur mainly on the base-rich parent materials such as drifts derived from limestone and basic igneous rocks, and drifts derived from Old Red Sandstone rocks along the Highland edge. On other parent materials, brown forest soils are rare, being confined to steep valley slopes at low altitudes. Brown magnesian soils and associated magnesian gleys are restricted to soils with a mainly ultrabasic or serpentinite component, as, for example, in the Cabrach area.

Noncalcareous, humic and peaty gleys, poorly or very poorly drained, are widespread but localized, occurring in receiving sites such as concave lower slopes and embayments in hill sides. The upper limit of these wet soils is often demarcated by a spring line. Peaty gleys are the most abundant type of gley, a reflection of the dominance of relatively high ground and acid parent materials throughout the region. Noncalcareous and humic gleys, largely confined to lower elevations, rapidly give way to peaty gleys with increase in altitude, the few areas above 300 metres being associated with the more base-rich parent materials. Wet soils are most common among the gently sloping hills of Angus, though gleys, especially peaty gleys, are considerably less extensive than those in upland areas of regions farther west.

Rankers and lithosols, normally related to outcropping rock or boulders, form a prominent but small proportion of the soils of the region. Most consistently associated with parent rocks such as quartzites, granites and serpentinites, they tend to become more common westwards and at higher elevations, the western Cairngorm Mountains being a major location.

Some of the larger expanses of heather moorland in Scotland are located in the North-East Grampian Highlands, indicative of the dominance of podzolic soils. These moors are managed for grouse-shooting, deer-stalking and rough grazing for sheep. Forestry plantations, together with some renowned natural Scots pine woods, are concentrated normally in valleys, especially of the Rivers Spey, Dee and Don, and on adjoining slopes, though in Glenbuchat a planting elevation of over 500 metres is attained. Intensive agriculture is restricted largely to sheltered valleys and pockets of more fertile soils, large tracts of land being unsuitable for cultivation due to climatic, slope or soil limitations. In upper Strathdon and in the Cabrach area, arable cultivation is practised at almost 450 metres, possibly the highest in Scotland.

THE NORTH-WEST GRAMPIAN HIGHLANDS

The North-West Grampian Highlands (Fig. 2) rise sharply from the Great Glen in the west and the River Spey in the east, stretching from the lowlands of the Moray Firth to the Loch Laggan axis in the south. They comprise the Monadhliath Mountains, mainly above 500 metres, and the Moray Firth Foothills; the latter contain the 305-metre peneplain with various monadnocks, for example, the Knock of Braemoray (455 metres) and the complex zone fringing the lowlands.

Apart from the massive corries of the Creag Meagaidh range and those west of Newtonmore, the region is one of featureless plateaux and gently rounded summits. Only the deeply incised valleys, such as those of the Rivers Tarff, Killin and Findhorn, relieve the monotony which is emphasized by the widespread blanket peat. From a summit plateau which varies mostly between 730 and 915 metres, the topography descends northwards by a series of steps; the derivation of these platforms from a former Highland peneplain created from the worn-down remnants of the ancient mountains of the Grampian Caledonides has been examined at length by numerous investigators; they include Geikie (1901), Peach and Horne (1930), Bremner (1942), Linton (1951), George (1965) and Sissons (1976). Regardless of the evolutionary hypotheses, the outcome has been the denudation of the Old Red Sandstone and later rocks from the pre-Cambrian Highland peneplain; remnants of the Old Red Sandstone basal conglomerates occur in the Findhorn basin. Within the exposed metamorphic rocks, zones of weakness along the south-west to north-east regional orientation have been exploited by the evolving drainage system. Thus the longitudinal development of the major rivers, the Nairn, Findhorn and Spey, has led to the capture and dismemberment of the original easterly, consequent streams by the constant adjustment to erosion.

With regard to the contemporary planation levels, Peach and Horne identified a High Plateau (610–915 metres) and an Intermediate Plateau (305 metres) whereas Fleet (1938) recognized the Grampian Main Surface (730–945 metres), the Grampian Lower Surface (450–640 metres) and a more restricted one, the Grampian Valley Bench (230–305 metres). Sölch (1936) outlined similar platforms.

Although the area is underlain largely by crystalline metamorphic rocks, there are associated igneous intrusions which post-date the regional metamorphism and foliation of the surrounding altered sediments (Fig. 3). The metasediments consist primarily of pelitic schists and gneisses, semi-pelitic schists and gneisses and striped schists formed from alternating bands of psammites and pelites. Other minor rocks include quartzites in the Dulnain valley, where soils of the Durnhill Association have been mapped, and associations of amphibolite, calc-silicate-granulites and crystalline marble which occur especially between Kincraig and Grantown-on-Spey. Apart from Laggan Hill near Dulnain where soils of the Deecastle Association have been mapped, these lithological varieties have only a local influence on the soils.

Originally, all these metamorphic rocks were assigned to the Moinian

24

Assemblage and although no stratigraphical subdivision was considered possible, the affinities with the Dalradian Assemblage of the pelites around Grantown and other areas were noted (Hinxman, 1915; Johnstone, 1966). Subsequently, the pelites and semi-pelites were identified as part of the lower Dalradian (Smith, 1968; Harris and Pitcher, 1975). Recent extensive radiometric measurements and surveys, however, have led Piasecki and van Breemen (1981, 1983) to question the Dalradian elements and to propose two structural-stratigraphic divisions for all the original Moinian rocks south-east of the Great Glen; they are the Central Highland Division and the Grampian Division. The former covers 1000 square kilometres between Loch Duntelchaig and Aviemore together with inliers at Kincraig and Laggan; it comprises a basement complex of psammites, semi-pelites and occasional quartzites, all of which are variably gneissic and migmatitic. The Grampian Division is a younger more extensive assemblage of finer-grained, flaggy psammites, semi-pelites, pelites and rarer calc-silicate-marbles. Although controversy surrounds the stratigraphy of these metasediments, the derived soils are broadly similar and all have been mapped within the Arkaig Association.

Several major intrusions of pink or red, medium- to coarse-grained biotite-granite are scattered across the region; they include the Ardclach, Monadhliath and Moy areas, each covering approximately 75 square kilometres. Smaller areas of dominantly red biotite-granite at Strathdearn, Corrieyairack and Alt Chrom also include biotite-hornblende-granites which are usually grey in colour. At Foyers, there are localized areas of more basic types including appinite and tonalite. Collectively, all these granites form the parent material of the Countesswells Association. Where the granites are interfelted with or include major blocks of the surrounding schist rock, for example, at the Foyers, Grantown-on-Spey and Alt Chrom intrusions, a parent material intermediate to those of the Arkaig and Countesswells Associations has been produced, that of the Aberlour Association. This is distinguished from the Arkaig and Strichen Associations by a reddish cast and a slightly gritty texture together with a noticeable content of granite stones.

Along the Great Glen between Loch Duntelchaig and Foyers, there is a narrow strip of Middle Old Red Sandstone conglomerates with subsidiary basal breccias and sandstones. The soils in this rock-dominated area are very stony and coarse textured and belong to Kessock Association.

Whereas preglacial planation surfaces dominate the physiography, glaciation has had a major influence in shaping local topography and producing soil parent materials. It is unlikely, however, that the Monadhliath Mountains ever developed more than a local accumulation of ice. Clear evidence, for example, the crag and tail ridges and associated overflow channels north-east of Carrbridge, indicates the passage of a major ice-sheet from the Rannoch basin along the Loch Laggan and Loch Ericht axes into Speyside and northwards. In the 305-metre peneplain, there are massive exposures 30 metres thick which show a red sandstone till overlain by a grey schist till; they suggest substantial oscillations between the Rannoch and Moray Firth ice-sheets. Deposits of till sometimes capped by moraine or fluvioglacial deposits and up to 30 metres thick overall are found also in the Monadhliath Mountains. They can be traced up to and across cols at 550 metres though most occupy embayments with gentle slopes below 400 metres; these benches probably equate with the Intermediate Plateau (Peach and Horne, *op cit.*) and the Grampian Valley Bench (Fleet, *op cit.*). Above these levels, erosion has been generally severe and on the summits most soils are formed from frost-shattered and weathered rock. In places, moraine or ablation till occurs as a veneer of gravelly loamy sands up to 1 metre thick.

Plate 6. *Near Carn nan tri-tighearnan (614 metres) on the Nairn-Findhorn watershed. The blanket peat of map unit 4e (Organic Soils, Class 7) is between 2 and 3 metres thick and shows an advanced stage of erosion. The common occurrence of* Racomitrium lanuginosum *and lichens reflects the drying out following the marked dissection by gullying. Sheet erosion accompanies the terminal phase.*

Blanket peat now dominates the Monadhliath plateaux and the 305-metre peneplain because of the combination of a rainfall between 1200 and 2000 millimetres, low temperatures and gentle slopes (Plate 6). Of the mineral soils which mainly flank the plateaux, the dominant parent materials are tills of the Aberlour, Arkaig and Countesswells Associations; they have sandy loam and loam textures with 10 to 15 per cent clay. Soils developed in thin drift on shattered rock usually have loamy sand or sandy loam textures with clay values dropping to around 4 per cent; a typical clay value for weathered schist is also around 4 per cent.

In contrast with the deeply dissected terrain to the south and west, there are limited glacially deepened valleys and breaches. Hummocky moraines are therefore restricted largely to the valleys and corrie floors along the southern boundary between Loch Spey and Newtonmore. The closely related, partially stratified, gravelly loamy sands (Dulsie Association) are, however, widespread below 300 metres in the Moray Firth Foothills and the basins of the Spey and Dulnain valleys. They form shallow mounds or, more commonly, a veneer up to 3 metres thick and are distinguished from fluvioglacial deposits by their poor sorting and constant silt values of around 20 per cent. Associated with the deglaciation are enormous spreads of fluvioglacial sands and gravels (Corby/Boyndie Associations) deposited mainly as pitted outwash plains around Lochindorb and in the Dulnain and Spey basins. Characteristically, they have less than 5 per cent combined silt and clay. They occur also in most valleys where they have been frequently reworked into flights of terraces — up to 11 terraces are found in the mid-Findhorn valley.

Acid coarse-textured parent materials and active leaching due to high rainfall, together with relatively low temperatures, have led to podzolization as the main soil-forming process. Except where very steep slopes induce colluviation, the podzolic major soil subgroups follow an altitudinal zonation. An iron pan above a friable B horizon is typical of all podzols below the subalpine soils zone and usually leads to surface-water gleying of the upper horizons on gentle slopes below spring lines and on steep slopes in the high rainfall areas. Between 500 and 700 metres, subalpine soils, mainly freely drained podzols and podzolic rankers, are found on the shallow summits protruding above the peat mantle; often they can be mapped only as a complex with the peat because of scale limitations. In severely eroded areas on the highest plateaux, the shallow soils are dominated by lithosols and rankers with peat restricted generally to cols. Elsewhere, at these altitudes and on exposed sites as low as 500 metres, alpine soils are common. They are normally freely drained podzols with a well-developed and organic-stained B horizon which has a very loose fabric presumably resulting from the annual freeze-thaw cycles. Under snow-bed vegetation, prominent H and E horizons may be found. Periglacial features, mainly terracettes, stone stripes and boulder lobes, are common throughout the zone.

Induration with a strong impermeable platy structure, and often accompanied by an iron pan or iron concretions on the upper surface, is widespread except in the alpine soils zone. Well-graded soils, for example, fluvioglacial sands and gravels, immature soils and soils subject to ground-water fluctuations are seldom indurated. Conversely, soils derived from tills and moraines usually have an indurated B horizon. In places the induration extends through several metres to bed-rock. Surface-water gleys often display a marked bisequal drainage regime with wet upper horizons resting directly upon the freely or imperfectly drained indurated subsoil; such profiles are regarded as fragogleys.

Peaty gleys are generally restricted to gentle and strong slopes below spring lines and are especially prominent where the till abuts the rock-controlled upper slopes of the rounded foothills. They become common towards the west and south-west as rainfall increases above 2000 millimetres. There, they are often a major component of complexes on rocky slopes and extend up to the alpine soils zone. Unlike gleys to the north and east, these gleys have common humus staining in the upper horizons and throughout the profile where induration is absent.

THE SOUTH-WEST GRAMPIAN HIGHLANDS

The area (Fig. 2) lies south of Loch Laggan, west of the main Perth/Inverness route through the Highlands, and extends as far south-east as that part of the Highland Boundary Fault running north-east from Loch Lomond to Birnam Hill, just south of Dunkeld. It includes some spectacular mountain scenery.

There is a close relationship between kinds of rock, structural pattern and topographic features. In general terms, the hard rocks form the highest hills and mountains, and the softer rocks are found in the valleys (Plate 7). The Highland Boundary Fault, extending north-eastwards from Loch Lomond along the north-west face of the Menteith Hills through Callander to Glen Artney, Comrie,

Logiealmond and Dunkeld, separates the South-West Grampian Highlands from the Midland Valley to the south. There are many summits in the mountain area approaching or exceeding 915 metres, whereas in the Midland Valley few summits exceed 610 metres. The forms of the hills are closely related to structure, their profiles being dependent on dip, strike or fracture.

The Dalradian rocks of the Highlands form rugged hills, a massif in relation to the softer rocks of the Midland Valley graben (Fig. 3). Within the regional landform, however, they suggest a dissected, stepped plateau whose summits may be regarded as the remnants of a tilted, eroded plateau. The dissection of the surface by river valleys, the differential erosion of rocks and the effects of glaciation have all combined to obscure the original form of the plateau surface but it probably sloped from the north-west to the south-east.

The highly dissected terrain of rugged pinnacles, crests and ridges of the South-West Grampian Highlands (Fig. 8) contrasts sharply with the more rounded summits of the high plateau further north and east. Resistant beds, such as quartzites, grits or massive gneisses, form lofty summits. If no marked structural planes are present, conical forms result, as in the quartzite mountain of

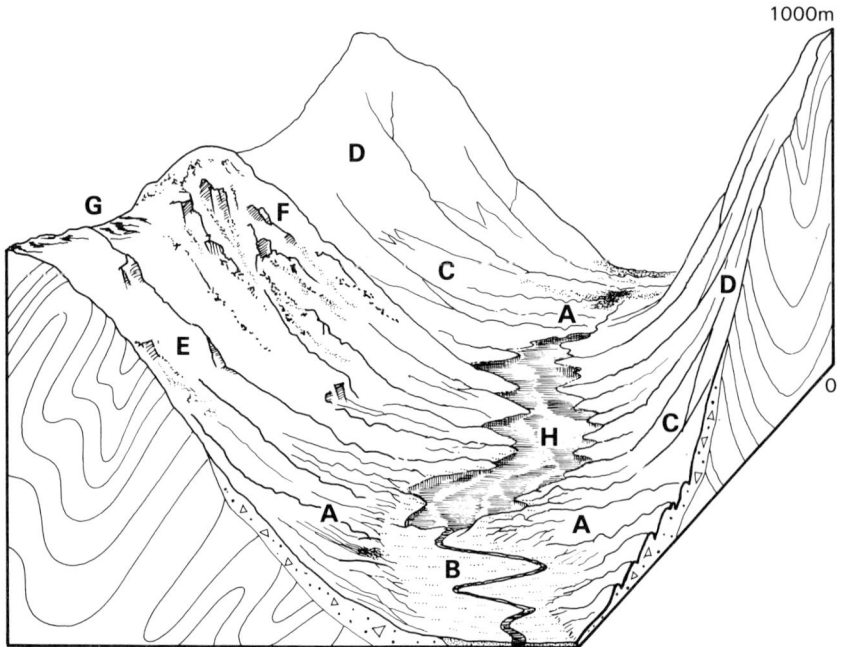

A   Hummocky valley moraine with peat flats
B   Alluvium and outwash, usually terraces and mounds
C   Steep gullied moraine with colluvial debris
D   Hill sides with steep and very steep slopes; non-rocky
E   Hill sides with steep and very steep slopes; moderately rocky
F   Hill sides with steep and very steep slopes; very rocky
G   Upper slopes, ridge crests and summits with alpine soils
H   Rock basin and loch

Figure 8. *Landforms of the South-West Grampian Highlands.*

Schiehallion. A series of metamorphosed grits, which forms a line of conspicuous mountains close to the Highland Boundary Fault, includes Ben Lomond, Ben Ledi and Ben Vorlich, whereas harder gneisses and slates form the summits of the highest ground in the Ben Lawers group of mountains. Between these resistant quartzites, grits, gneisses and hard slates, belts of weaker strata such as limestones, schists, phyllites and softer slates have been excavated into valleys. Some of the intrusive igneous rocks, including diorite and granite, also form prominent summits, such as Ben Chonzie near Crieff, but these are not common in the South-West Grampian Highlands.

Together with the rest of Scotland, the South-West Grampian Highlands were blanketed during the Pleistocene Period by ice which did not melt finally until some 10 000 years ago. Signs of the glaciation which greatly modified the preglacial landscape are thus very fresh and impressive and are seen in a great variety of features. In general, the ice-flow followed the pre-existing valleys in a south-east and east direction from the Highlands to the North Sea basin. The mountains and associated valleys were subjected to intense erosion and display an abundance of ice-moulded profiles. The low ground received ice-transported debris, glacial drift, which veneered and obscured the solid rock on which it came to rest.

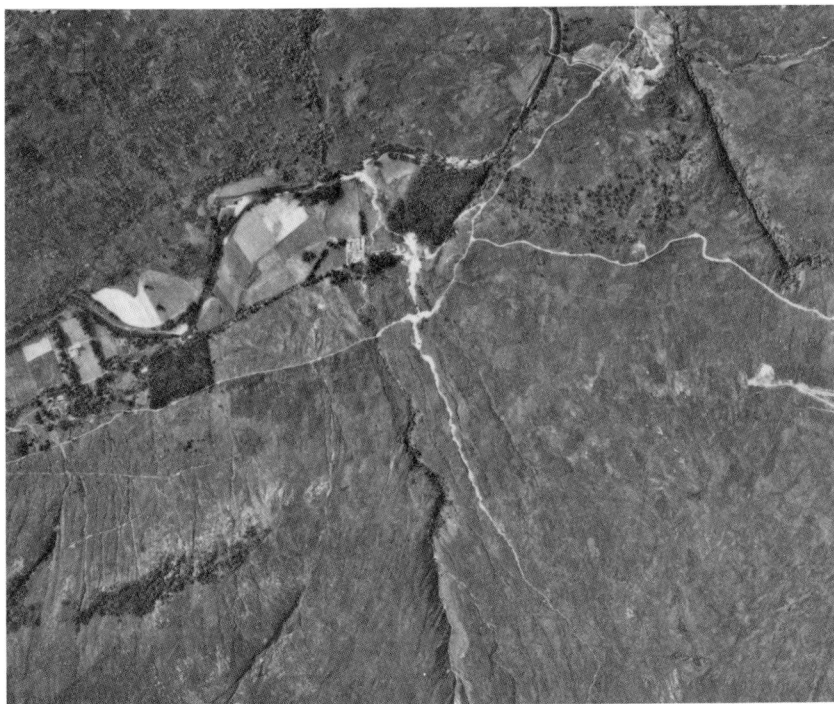

Plate 7. *A vertical air photograph (original scale 1:27 000) of upper Glen Lyon. Along the narrow valley floor, the humus-iron podzols and alluvial soils of map unit 98 (Corby Association, Class 5) provide the only good quality grassland. The very steep valley sides with the brown forest soils, humus-iron podzols and gleys of map unit 505 (Strichen Association, Class 6) provide only moderate rough grazing. On the hill and moor beyond, the peat, peaty gleys and peaty podzols of map unit 23 (Arkaig Association, Class 6) provide only poor rough grazing.* Ordnance Survey photograph; Crown copyright.

29

Glacial erosion has been spectacular in the Highlands. Steep-walled corries with rugged slopes and swathes of late- and post-glacial apron scree are not as common in the South-West Grampian Highlands as they are further north, but good examples exist on the flanks of Ben Vorlich, Ben Ledi, Stuc à Chroin and the Ben Lawers range of mountains. The heads of many valleys are deeply scalloped, some with the near-vertical walls of developing corries and with an associated contemporary, but misfit, drainage system.

Deeply gouged valleys occur throughout the area and are the major elements contributing to the rugged appearance of the mountains. In their extreme forms, they are related to the excavation of glacial rock basins whose floors are now occupied by elongated lochs such as Lochs Katrine, Voil, Earn, Tay and Rannoch. A common feature of the lochs is the formation of deltas where inflowing streams are reduced in velocity and shed their suspended material on entering the relatively still loch waters. Frequently, these alluvial deposits provide the only arable land in such glaciated valleys.

At the close of the Ice Age, the ice melted first on the lower ground, its front subsequently receding and breaking into many tongues which filled the upper reaches of the valleys. On melting, it deposited vast quantities of morainic material, boulder clay and water-borne sand and gravel in valley bottoms and on lower ground, Much of this debris forms the parent material of present-day soils.

The distribution of the major soil subgroups is very much related to climate, topography and parent materials in most areas, but especially in the area of the South-West Grampian Highlands with its dominantly east–west valley system. In most main valleys the soils and vegetation on slopes with a northerly aspect contrast sharply with those on the slopes of the opposing, southerly aspect of the same valley, the latter having a better climate. Brown forest soils are common on the lower, sheltered slopes of the main valleys, together with gleys in flushed and poorly drained sites. The former soils extend further up slopes which have more southerly and westerly aspects than those with northerly and easterly aspects. This is especially evident where soil parent materials are enriched with less acid rocks such as hornblende-schist and impure limestone. The brown soils are succeeded on the steeper, upper slopes of the valley sides by humus-iron podzols or by gleys in the wet sites and by peaty podzols on the gentler slopes at the highest level. The increased rainfall of the higher ground, coupled with the more severe climate, have favoured the formation of peaty-topped soils and the accumulation of hill peat, especially on the gentler slopes. In the very wet, western areas, however, wet peaty soils and peat occur on a variety of slopes, some of which are quite steep. The very exposed mountain ridges have subalpine soils on their sheltered flanks and alpine soils on their summits, interspersed with lithosols and rankers where rock outcrops are frequent. Bare rock and scree are present on some of the very rugged ridges and summits but are less common than further north.

THE NORTHERN HIGHLANDS

Shown on Fig 2, the Northern Highlands occupy the north-western corner of Eastern Scotland (Sheet 5) and are bounded to the east by the Moray Firth Lowlands and by Glen Mor, the Great Glen; the region is part of the much larger area of mainland Scotland west of the Great Glen, the geology and physical features of which are described by Phemister (1960). The rocks of this larger area are mainly of pre-Cambrian age with some Palaeozoic rocks and, very locally on the west coast, some Mesozoic rocks and some volcanic rocks of the early Tertiary

Period. After the volcanic episode, the area formed part of a great continental block, which was reduced by denudation to a peneplain and later elevated to a high tableland with a watershed aligned in a south-west to north-east direction. It now comprises a dissected plateau, highest in the south and sloping northwards. The initiation and development of the drainage system has been the focus for many hypotheses which are reviewed by Sissons (1976). Subsequent drainage towards the north-east and south-east was developed along the strike of less resistant rocks and along fault lines. Eastward flowing, consequent rivers include the River Conon and in Glen Urquhart, the River Enrick. Many of the larger valleys are steep-sided and comprise a striking element of the scenery.

The rocks of the region are mainly those of the Moine Assemblage, roughly foliated quartz-feldspar-biotite-granulites with subsidiary bands of mica-schists, semi-pelitic schists and mixed schists; the rocks are cut by occasional dykes of epidiorite and hornblende-schist (Fig. 3). One band of the schists forms the highest ground of the region, Ben Wyvis (1046 metres), which has spectacular corries on its eastern and northern flanks. Rocks of the Moine are the source for parent materials of the Arkaig Association. The area is predominantly foothill and upland with elevations ranging from 200 to 800 metres.

Outcrops of Lower and Middle Old Red Sandstone rocks, mainly conglomerates and sandstones, form a discontinuous run along the eastern border of the region. These outliers are the remnants of a formerly more extensive cover and some, for example, the one north of Drumnadrochit, are associated with faults.

The conglomerates are resistant to weathering and locally form inselberg-like hills such as Cùl Mór, Meall Mór and Meall nan Caorach; Meall Fuar-Mhonaidh (696 metres) south-west of Drumnadrochit, forms the highest ground in the Old Red Sandstone. These conglomerates are the source rocks for the parent materials of the Kessock Association and the sandstones are the source rocks of the Sabhail/Mounteagle Associations. North of Dingwall, the parent materials of the Braemore Association are derived from mudstones, shales and fine sandstones. As with that of the Moine Assemblage, the area occupied by Old Red Sandstone rocks is mainly foothill and upland.

West of Loch Glass, igneous rocks include the edge of the Carn Chuinneag–Inchbae intrusion, a foliated coarse biotite-granite or augen gneiss. There is also an outcrop of granite on the steep rocky slopes above Loch Ness, north-east of Abriachan. These intrusions provide the source rocks for the parent materials of the Countesswells Association. Finally, the oldest rocks of the region are the inliers of the Lewisian ultrabasic rocks and gneiss west of Drumnadrochit, but the areas of derived soils are too small to map as separate soil associations.

During the Pleistocene Period, ice and meltwater modified the landscape both by erosion and deposition. Examples of erosion include the over-deepened Lochs Ness and Garve and the numerous meltwater channels including the Raven Rock channel, east of Loch Garve, and the gorge cut in conglomerate in the Beauly valley at Kilmorack. Striae and erratics of Inchbae granite provide evidence of an easterly ice movement (Small and Smith, 1971). In many areas, particularly in the uplands and mountains, rock outcrops are present and deposition is limited to a locally derived, stony drift less than 1 metre thick. Minor rock outcrops of Old Red Sandstone conglomerate often differ from those of the Moine in being smoother in outline and thus less obvious. Although Ben Wyvis has prominent corries and other very rocky areas, it nevertheless is unusual in having an only slightly rocky summit and smooth westerly and southerly slopes. Most Moinian hills in the region are moderately and very rocky. Thicker drifts occur in the lowlands and foothills. They are usually coarse or moderately coarse in texture,

31

mainly sandy loams and loamy sands, but compact tills, tending to be finer in texture, occur locally. These include sandy loams to clay loams in Strath Sgitheach north of Dingwall, where shales contribute to the till, and in Glen Convinth south of Beauly. Further evidence of an easterly ice movement is provided by the presence of Old Red Sandstone material in the drift overlying Moine rocks in The Aird, and by Moine material in similar drift overlying Old Red Sandstone rocks west of Beauly. The mixed drifts are the parent materials for soils of the Orton Association. Fluvioglacial gravels occur in the valleys and form soils of the Corby Association. The largest area is in the valley of the River Ness where there are prominent kame terraces, eskers and kettleholes. Similar landforms occur downstream from Loch Glass.

Most of the soils of the region are podzols with gleys and peat confined to receiving sites in basins and flats, to slopes below spring lines, and to some of the gentle and strong slopes in the uplands. As would be expected, the proportion of poorly drained soils tends to rise as rainfall increases. For example, the soils of *map unit 28* on the strongly undulating plateau north of Glen Urquhart are mainly peaty podzols, humus-iron podzols, peat and peaty gleys, whereas those of *map unit 29* south of the glen, on similar topography, albeit at rather higher altitudes, are mainly peaty gleys and peat. The rainfall is 1100–1200 millimetres in the northern area and 1200–1600 millimetres in the southern one. Glen Urquhart itself is typical of other straths in the region in having a high proportion of freely drained soils, mainly humus-iron podzols, on the strong and steep slopes of its valley sides. Most of the cultivated land in the region is in the eastern foothills and valleys where the cover is thick, the soils being mainly cultivated humus-iron podzols. Subalpine and alpine soils, mainly freely drained podzols, are confined almost entirely to the Ben Wyvis area. There, the gentler, concave slopes of the foothills are mostly peat-covered with subalpine soils on the steep slopes above about 600 metres and alpine soils above about 750 metres on the upper convex slopes and on the summit.

CLIMATE

The climate of Eastern Scotland is characterized by its considerable variability within the limits set by latitude and the maritime influence (Fig. 9); in a few miles it changes from mild oceanic conditions to a harsh subarctic environment where the tree line barely exceeds 500 metres. Rainfall (Fig. 10), in particular, illustrates the range, varying from 600 millimetres on the eastern coast to over 3200 millimetres in the mountainous south-west. Nevertheless, the climate is related broadly to relief and is governed by a succession of cyclonic disturbances. These flow across the region from the relatively warm waters of the Atlantic except when high pressure systems are centred over northern Europe. When the boreal anticyclones occur in winter, the main airstream is often a cold north-easterly which brings Maritime Polar or Continental Polar air to the region. Such conditions often bring blizzards and deep snow to the exposed North-East Lowlands.

Although a daily mean temperature of 5.6°C has been accepted generally as the threshold for active plant growth, there is evidence that the figure may be too high. However, Birse and Dry (1970) used this baseline to calculate the accumulated temperatures in day-degrees Celsius and their simplified results are shown in Fig. 9. In general, the sheltered Moray Firth and Central Lowlands are warm (more than 1375 day °C); they benefit from the föhn effect created by the mountain mass to the west. The adjacent foothills and the lower slopes of isolated hills, for example, the Sidlaw Hills, are only fairly warm, the lower temperatures reflecting the increased altitude. Also classed as fairly warm are the Howe of the Mearns and the North-East Lowlands, both areas being affected by the cooling influence of the North Sea. Of the remainder of Eastern Scotland, mainly the Grampian Highlands, approximately half lies between 400 and 800 metres and is classed as cool (825–1100 day °C). Above 800 metres, the mountains are cold (550–825 day °C) with the very exposed, highest plateaux and summits being very cold. The growing season along the coastal strip decreases northwards from 245 days at St Andrews to around 235 days at Forres. As altitude increases, the number of days is reduced by approximately 1 day for every 8 or 9 metres, the reduction being qualified by distance from the sea, aspect, exposure and dates of the first and last frosts. Thus at Craibstone (91 metres), with an annual average daily mean temperature of 7.8°C, the days total 222; other examples are, Logie Coldstone (185 metres) 7.3°C, 207 days and Braemar (338 metres) 6.4°C, 190 days. Further inland and at higher altitudes, the mean annual temperature at Coire Cas Sheiling (762 metres) is 4.7°C and at Cairngorm (1090 metres) is 2.3 °C.

Annual average hours of bright sunshine in the lowlands are amongst the highest in Scotland and range from 1347 at Forres to 1505 hours at Arbroath. These figures are largely due to the föhn effect breaking up the cloud in the coastal areas. Elevation and aspect obviously affect the amount of sunshine, for example, in Morayshire in mid winter a northern hill side of 8° elevation would receive no sunshine (Ross, 1976). The cloudier conditions in the hills are illustrated by the total of 1130 hours at Glenmore (341 metres) and 1118 hours at Braemar (338 metres).

Because of the ameliorating influence of the North Sea, the lowlands are not subject to prolonged frost. Birse and Robertson (1970) used accumulated air frost (day-degrees below 0°C) to demonstrate that much of the coastal zone, apart from the very exposed headlands, has fairly mild winters (20–50 day-degrees). The bulk of the lowlands, however, is subject to moderate winters (50–110 day-degrees). Agronomically, the date and incidence of frost are very important with ground frost often occurring twice as frequently as air frost. Within the lowlands, the range of average number of days with air frost includes 46 at Arbroath (30 metres altitude, 101 days ground frost) and 88 at Blairgowrie (61 metres altitude, 141 days ground frost). The duration and intensity of frost in the foothills and mountains are very dependent upon exposure and altitude though on inland sites ground frost can be recorded in almost any month. Days of air frost range from 113 at Grantown-on-Spey (229 metres) to 196 days at Cairngorm (1090 metres). During spells of high pressure with no cloud and little wind, cold air drainage and frost hollow effect, especially in the valleys, lead to very low temperatures; in upper Speyside and Deeside the temperature may plunge to below −20°C.

Much of the rainfall is derived from the westerlies and is orographic with the highest precipitation and number of rainy days occurring in the extreme south-west (Fig. 10). Consequently, the eastern areas are considerably drier; they derive much of their precipitation from winds tracking across the North Sea. The lowlands, apart from the Forth valley, receive mainly less than 900 millimetres and the coastal fringe less than 600 millimetres; the isolated Ochil Hills and Gargunnock Hills have between 1600 and 1800 millimetres. Even the highest hills in the region, the Cairngorm Mountains, receive substantially less rain (1600–2000 millimetres) than that (2000–3200 millimetres) which falls on the high peaks of the South-West Grampian Highlands. With regard to the seasonal

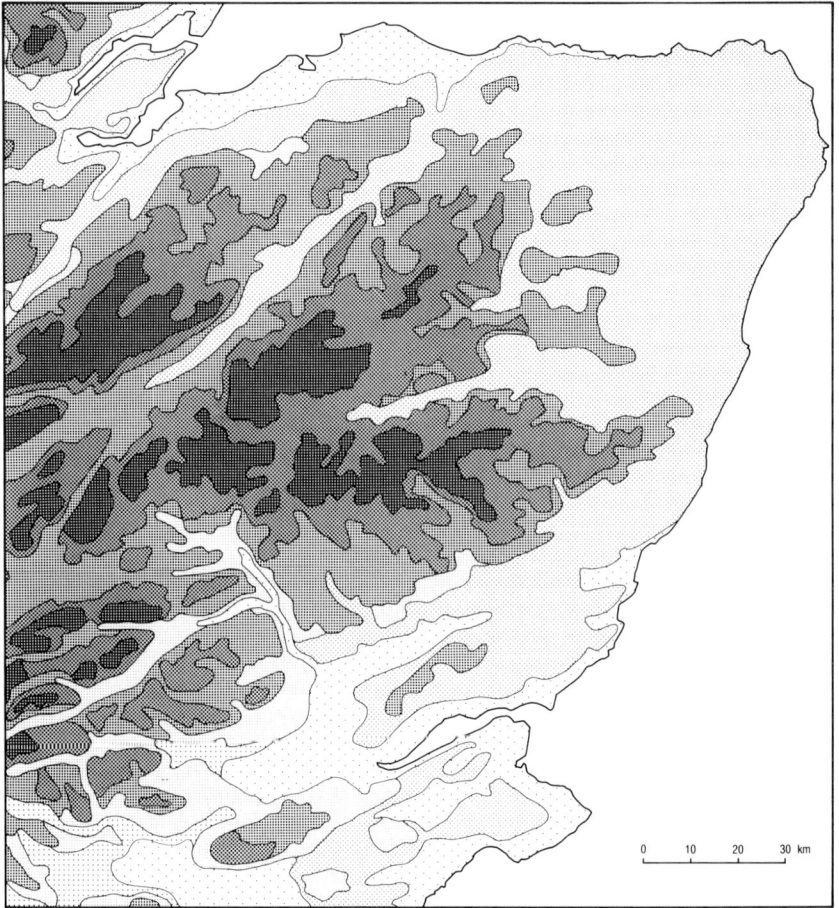

Figure 9. *Climate regions.*

distribution, there are similarities throughout the whole area, the driest months usually being in spring and early summer, with a secondary minimum in August and September, July being decidely wetter. The intensity of rainfall is a peculiar characteristic of the Cairngorm Mountains and, more especially, the Mona-dhliath Mountains where frequent devastating floods have been recorded in the rivers draining towards the Moray Firth. Such prolonged heavy rainfall appears to result from slow-moving depressions in the North Sea and the coincidence of orographic uplift with the convergence of surface air streams into constricted valleys.

Along with Shetland, Orkney and Caithness, the North-East Lowlands and the North-East Grampian Highlands have the greatest number of days when snow falls; all these areas are completely open to the snow-bearing winds from the north and north-east. The number of days, during which snow lies as opposed to falls, varies according to altitude and degree of shelter. It increases rapidly inland at approximately one day more per year with snow falling for every 15 metres of

warm and moderately dry

warm and wet

fairly warm and moderately dry

fairly warm and wet

cool and wet

cold and wet

very cold and wet

Accumulated Temperature Divisions		Potential Water Deficit Divisions	
RANGE (day °C)	DESCRIPTION	RANGE (mm)	DESCRIPTION
›1375	warm	›75	dry
1100-1375	fairly warm	25-75	moderately dry
825-1100	cool	0-25	wet
550-825	cold		
0-550	very cold		

Modified from Birse and Dry (1970)

elevation above 60 metres (Ross, 1976). In the North-East Lowlands, the average number of days with snow lying is around 30 whereas in the eastern Central Lowlands the number varies from 15 at Cupar to 23 in Strathmore. By contrast the observed snow cover in the Cairngorm area ranges from 60 days at 300 metres to more than 200 days at 1220 metres (Manley, 1971), where snowfall is likely to exceed 100 days (Green, 1975).

Excluding the upper Forth valley, humidity in the lowlands varies from 50–70 per cent in spring to 60–80 per cent in summer and rises upwards of 80 per cent in winter. The summer figures largely reflect the haar, or advection fog, and the increased rainfall associated with the east and south-east winds; the spring figures probably reflect the influence of the föhn conditions. Average potential evapo-transpiration values during summer for the coastal areas of Moray and Fife are around 425 millimetres whereas according to Birse (1971) much of the South-West Grampian Highlands and the highest summits of the Monadhliath and the Cairngorm Mountains have summer rainfall exceeding evapo-transpiration by

Figure 10. *Rainfall (average annual, mm).*

more than 500 millimetres. Average annual potential water deficits range from 40 millimetres at Aberdeen to 80–90 millimetres between Peterhead and Inverness; from 31 millimetres at Kingussie, values drop rapidly to 22 millimetres at Loch Rannoch to 0 millimetres at Glen Lyon and Loch Katrine.

According to the exposure map produced by Birse and Robertson (1970), much of the lowlands is moderately exposed (average wind speed 2.6–4.4 metres per second) with the minor Moray Firth Lowlands being sheltered (less than 2.6 metres per second). The Grampian Highland foothills are mainly exposed (4.4–6.2 metres per second) whereas the higher uplands are very exposed (6.2–8.0 metres per second); the highest peaks and plateaux are severely exposed (greater than 8 metres per second). Because the wind speed of the prevailing south-westerlies is checked by the land surface, gales (Force 8 or 22 metres per second) are not frequent in the lowlands. They vary from an average 12.7 days in the Moray Firth Lowlands to 7 days in the Central Lowlands. Records for the mountains are sparse but recent, incomplete records on Cairngorm suggest 100 days, mostly in the winter months; the figure is probably conservative. In March 1967, the average wind speed on Cairngorm for the whole month was over 15.5 metres per second.

VEGETATION

In the following account, the distribution of the plant communities and their relationship with the soils of Eastern Scotland (Sheet 5) are briefly discussed. The common names quoted for these communities in the text, both here and under each soil association or map unit description (Chapter 2), are based on the vegetation field units used in the 1:250 000 survey. These units are listed and described in Handbook 8. Individual species names follow those of Clapham, Tutin and Warburg (1962) for vascular plants as do the bulk of the common names, those of Smith (1978) for mosses and those of James (1965) for lichens.

The classification of the plant communities in phytosociological terms is quoted in brackets after each community name and follows that of Birse and Robertson (1976) and Birse (1980, 1982). When a community is firmly established as an association, it is put in the Latin form (-etum), but when there is some doubt as to the validity of the association, it is named by one or two plant species followed by the term 'Association'. When there are insufficient records to establish an association, the vegetation is again named by one or two plant species but with the term 'Community' following.

The vegetation that is found within any particular area can be considered as a natural expression of its environment. Thus, the diversity of the country rocks of a region and their derived parent materials, the genetic soil groups, the landforms, the climate and the land use, will all be reflected to some extent in the range and form of plant communities that are found there. Eastern Scotland is the most diverse in these respects of the seven survey areas established for the production of the 1:250 000 soil and land capability for agriculture maps and two main zones of variation may be recognized. The first extends broadly from east to west and is an expression of the amount of rainfall with the dry lowland forms of vegetation more widespread in the east and flushed forms to the west. The second is an expression of exposure and altitude whereby 'northern' forms of communities are found on more exposed sites to the west, and alpine or oroarctic plant communities replace those of the foothills and uplands on the windswept ridges and summits of the higher hills and mountains. Parent materials range from drifts derived from limestone and ultrabasic rocks with the soils supporting species-rich communities to those derived from granites and quartzites with species-poor vegetation. The soils developed on these parent materials also exert an important influence. The brown forest soils and podzols of the steep valley sides and lower hill slopes carry grassland and moorland communities of significant grazing value and are the sites of most of the fragmentary broadleaved woodlands, whereas the peaty soils and peat of the uplands support only moorland and blanket bog communities of little agricultural worth. Land management has improved hill pastures within easy reach of the farms and crofts where landform permits, has created rough grassland from moorland on the foothills by fencing and grazing and has maintained a mosaic of moorland vegetation of varying age by periodic, controlled burning.

Grassland

The main areas of arable agriculture are concentrated in the north and east on the coarse-textured coastal soils of the Moray Firth, Angus and Fife and on the till plains of Buchan, the Howe of the Mearns, the Vale of Strathmore and the lowlands of Fife and Kinross. In the south-west, the principal crop grown on the moderately fine- and fine-textured lowland soils under high rainfall is grass in the form of ley, long ley or permanent pasture (Lolio-Cynosuretum). Where such pastures have been allowed to deteriorate through undergrazing or loss of fertility, the replacement community of meadow-grass–bent pasture (the *Galium saxatile–Poa pratensis* Community) may become established. This vegetation may be formed also from semi-natural rough grasslands such as bent–fescue (Achilleo-Festucetum tenuifoliae) or common white bent (part of Junco squarrosi-Festucetum tenuifoliae) as a result of heavy grazing and dunging. Frequently, it is found on dry ridges or around rock outcrops on the lower and mid-hill slopes where animals tend to 'lie-up' for shelter. It is also the characteristic vegetation of immature freely and imperfectly drained alluvial soils in sheltered valleys where grazing animals congregate. Poorly drained, peaty-topped soils of these sites carry soft rush pasture (the *Ranunculus repens–Juncus effusus* Community) or, less commonly, tussock-grass pasture (the *Deschampsia cespitosa* Community) and these communities are to be found also on low base status noncalcareous gleys, peaty gleys and flushed peat of the lower hill slopes. Another rush-dominated community of wet, undrained lowland and foothill soils is sharp-flowered rush pasture (Potentillo-Juncetum acutiflori) which may be present in its species-poor form on the same soils as mentioned for the soft rush pasture or as its species-rich form on better base status noncalcareous and humic gleys. This vegetation is generally confined to wet channels and depressions except in the south-west where it is extensive, spreading out over the flushed hill slopes. Many other communities of wet alluvial flats and basins, flushed slopes and drainage channels may be found throughout the landscape, but these are usually of very limited areal extent in all but the higher rainfall areas of the west. The presence of any one of these communities is dependent on the base status and nutrient level of the flush water and so the more eutrophic vegetation will tend to occur on the receiving sites of the lowlands and the dystrophic forms on the peaty flushes of the higher hill slopes. Thus, at the dystrophic end of the scale, bog moss water track (the *Juncus effusus–Sphagnum recurvum* Community) is found on peaty gleys and peat associated with the drainage water from peat bogs as are common sedge flushes (*Carex nigra* communities) and star sedge mire (Caricetum echinato-paniceae). More mesotrophic vegetation, such as few-flowered spike-rush mire (Carici dioici-Eleocharitetum quinqueflorae) and flea-sedge mire (Caricetum hostiano-pulicaris), is confined to low base status noncalcareous gleys, peaty gleys and peat in flushed channels and depressions, whereas the moderate to high base status noncalcareous gleys, humic gleys and eutrophic peats of lowland swamps and marshes support a wide range of communities which includes meadow-sweet meadow (Valeriano-Filipenduletum), marsh marigold meadow (the *Caltha palustris* Community), reed grass swamp (Phalaridetum arundinaceae) and reed swamp (Phragmitetum communis).

Semi-natural rough grasslands, as mentioned above, often have been created over a long period by land management and are usually to be found on the hills surrounding farms or crofts. Others have been formed by generations of grazing animals in areas of natural shelter such as narrow valleys and corries. Although grassland communities are found throughout the map area, the greatest concentration lies on the Perthshire hills, the Ochil and the Sidlaw Hills, the green glens of Angus and the basalt uplands of Fife and Kinross. Bent–fescue grassland (Achilleo-Festucetum tenuifoliae) is the characteristic community of lowland hill and valley slopes on brown forest soils and, less commonly, on humus-iron podzols. It represents the best of the natural hill grazings and is usually present in its acid or typical form. A dense cover of bracken may dominate the vegetation, especially on the steeper, uneven, rocky slopes and in its extreme form can greatly reduce the productivity of the underlying herbage. The canopy, however, is usually more open and the community provides both grazing and shelter for young animals. Herb-rich bent–fescue grassland is much more localized, being confined to the steep slopes of mounds and valley sides on brown forest soils and brown rankers developed on the more base-rich drifts such as those derived from limestones, gabbros, basalts and intermediate lavas. Heath rush–fescue grassland (Junco squarrosi-Festucetum tenuifoliae) is another important element of the hill grazings but at a much lower level of productivity. The drier form, common white bent grassland, which is found on peaty podzols, humus-iron podzols and, less commonly, on peaty gleys, occurs extensively on the upper slopes of the Ochil Hills. Heath grass–white bent grassland is a herb-rich form which is confined to hill slopes subjected to a degree of flushing mainly in the south-west. The associated soils include brown forest soils with gleying, peaty podzols, peaty, humic and noncalcareous gleys and peat. The wettest form of the association, flying bent grassland, is extensive only on hills in the west on peaty gleys, blanket peat and some peaty podzols, being confined elsewhere to local hillside flushes and channels.

Mention has already been made of the influence of parent materials on the vegetation and a number of grassland communities owe their existence to the presence of soils derived from basic or ultrabasic drifts. Serpentine tussock-grass grassland (the *Helictotrichon pratense–Deschampsia cespitosa* Association), for example, is found only on the brown magnesian soils and magnesian gleys of the flushed ultrabasic hills near Strathdon. In Glen Avon and around Blair Atholl, the limestone community of rockrose–fescue grassland (the *Galium sterneri–Helictotrichon pratense* Community) colonizes the shallow brown forest soils and brown calcareous soils associated with rock outcrops and colluvial slopes, whereas the flushed form with hair sedge (*Carex capillaris*) occurs in the calcareous flushes. The rare false sedge (*Kobresia simpliciuscula*) is sometimes found in the latter sites. Brown lithosols and brown forest soils on the basalt and intermediate lava hills in the south support crested hair-grass grassland (the *Galium verum–Koeleria cristata* Community). More generally, on hill slopes mainly in the south-west above the zone of peaty soils, the combination of grazing pressure, high rainfall and base-rich rocks within the Dalradian schists and slates has resulted in a dominantly green landscape. The principal community there is species-rich upland bent–fescue grassland (part of Achilleo-Festucetum tenuifoliae), characterized by the presence of alpine bistort (*Polygonum viviparum*), alpine lady's-mantle (*Alchemilla alpina*) and other *Alchemilla* species, on flushed brown soils, noncalcareous gleys and humic gleys. White bent–tussock-grass grassland (the *Cirsium palustre–Nardus stricta* Community) is present on these flushed hill slopes though it is found elsewhere on similar sites but on a wider range of parent materials. In the north-east, these materials are derived partly, or mainly, from basic igneous rocks.

Moorland

The characteristic vegetation of the main massif of the eastern Highlands is that of moorland dominated by heather (*Calluna vulgaris*) where the chief management practice is periodic burning to maintain rough grazings for sheep, grouse and deer. Much of the hill land in the east and north-east is at altitudes of 300 metres or more where the climate is least influenced by proximity to the sea and the typical association on the better, drained soils of the foothills and uplands is that of boreal heather moor (Vaccinio-Ericetum cinereae). The dry form on humus-iron and peaty podzols and the moist form on peaty podzols and peaty gleys are both widespread and extensive, but the herb-rich form is much more localized, being confined to brown forest soils, brown magnesian soils and, less commonly, humus-iron podzols on the steep slopes of mounds, valley sides and foothills. A transitional zone often exists between the soils of the uplands and those of the exposed hill and mountain tops and this may be reflected in the vegetation by the presence of lichen-rich boreal heather moor in which the dwarf shrubs exhibit the initial effects of windcutting.

Outwith the periphery of the Grampian Mountains, boreal heather moor is replaced by its more oceanic counterpart, Altantic heather moor (Carici binervis-Ericetum cinereae). Only the dry form on humus-iron and peaty podzols and the moist form on peaty podzols and peaty gleys are at all common.

Bog heather moor (Narthecio-Ericetum tetralicis), the vegetation of peaty gleys and shallow peat associated with the gentle slopes bordering peat basins in the east, becomes more widespread on the blanket peat to the west under higher rainfall. A northern form, characterized by the presence of woolly fringe-moss (*Racomitrium lanuginosum*) and the lichens *Cladonia arbuscula* and *C. uncialis,* occurs chiefly in the north and west on the more exposed sites.

Blanket bog (Erico-Sphagnetum papillosi) is the most extensive of the moorland vegetation on level ground or gentle slopes where blanket peat has developed. Typical or lowland blanket bog vegetation and its northern or exposed form are both widespread, as is the upland form with crowberry (*Empetrum nigrum*) and hooked moss (*Rhytidiadelphus loreus*). An outstanding example of a lowland bog is Flanders Moss to the west of Stirling. Dwarf shrubs are less dominant on the blanket peat of the west where flying bent and cotton-grass bog colonize the flush channels and slopes. This change in dominance takes place along an imaginary line drawn approximately from Fort Augustus, Loch Laggan, Kinloch Rannoch, Loch Tay and Loch Earn to Loch Venachar. On the higher hills and mountains, the association is replaced by that of mountain blanket bog (Rhytidiadelpho-Sphagnetum fusci) in which cloudberry (*Rubus chamaemorus*) is typically present. A community derived from blanket bog or bog heather moor by persistent burning and grazing – deer-grass moor (the *Trichophorum germanicum–Calluna vulgaris* Association) – is sometimes found on the hills in the north. The characteristic elements of blanket bog – tussocks of cotton-grass and cushions of bog moss (*Sphagnum* species) – are suppressed and the vegetation is dominated by a mixture of heather (*Calluna vulgaris*), deer-grass (*Trichophorum cespitosum*), bog heather (*Erica tetalix*) and flying bent (*Molinia caerulea*). A northern or exposed form may be found and a closely related community on blanket peat with dwarf birch (*Betula nana*) and juniper (*Juniperus communis nana*) occurs in the Cairngorm Mountains.

Oroarctic communities

The subalpine and alpine soils of the high hills and mountains support a number of plant communities that reflect the influence of the hostile environment of infertile, often skeletal, soils, extreme exposure and severe winters. The most widespread association of smooth, rounded summits, flushed slopes and depressions which hold a covering of snow during the winter is the stiff sedge–fescue grassland (the *Carex bigelowii–Festuca vivipara* Association) which may be dominated by white bent (*Nardus stricta*), heath rush (*Juncus squarrosus*)

40

or stiff sedge (*Carex bigelowii*). The form with stiff sedge is usually extensive on the better-drained alpine soils of smooth summits such as Glas Maol and mountain heath rush grassland is probably a pioneer community of old eroded surfaces on soils with a peaty surface horizon. Mountain white bent grassland — much the most widespread and extensive of the forms — occurs in depressions and on concave slopes on high-level gleys that are seasonally waterlogged and on podzols. It is found also recolonizing exposed mineral surfaces within areas of hagged peat. The association provides the bulk of the high-level summer grazings. Alpine clubmoss snow-bed (the *Lycopodium alpinum–Nardus stricta* Community) is a local white bent-dominated community of late snow-lie hollows, especially in the Cairngorms, and bog whortleberry heath (the *Racomitrium lanuginosum– Vaccinium uliginosum* Association) may be found on similar sites. Blaeberry heath (the *Rhytidiadelphus loreus–Vaccinium myrtillus* Community) is an azonal community which is common on steep, upper hill slopes and corrie walls, especially on stabilized scree, on a wide range of soils including brown forest soils, podzols, peaty podzols and redistributed peat. The most exposed ridges and summits, where there is only light or moderate snow cover because of strong winds, usually carry a wind-cut mosaic of fescue–woolly fringe-moss heath (Festuco-Rhacomitrietum lanuginosi) and alpine azalea–lichen heath (Alectorio-Callunetum vulgaris) on freely, or occasionally imperfectly drained alpine soils or rankers. Fescue–woolly fringe-moss heath has a hyperoceanic distribution more westerly than that of the latter association, however, and is often absent from the hills in the east and south. At higher levels, three-leaved rush heath (Cladonio-Juncetum trifidi) is an open vegetation of alpine soils on the Cairngorm plateau, particularly on Ben Avon, and it also occurs on the exposed ridges of Ben Lawers.

Scrub and woodland

Coniferous plantations are the commonest form of woodland within the region and their distribution and extent are more fully described in the chapter on land evaluation. Many of these plantations have been in existence for some considerable time, especially pine plantations (the *Erica cinerea–Pinus sylvestris* plantation) in the north-east where the vegetation beneath the canopy bears a close resemblance to that of native pinewood, containing species such as lady's tresses (*Goodyera repens*) and lesser twayblade (*Listera cordata*) in a thick carpet of mosses. The soils are usually podzols. Further south in Angus and Fife, pine has often been planted on the cultivated soils of former agricultural land and this is reflected in a dense undergrowth of bramble (*Rubus fruticosus*) and ferns such as broad buckler-fern (*Dryopteris dilatata*), male fern (*D. filix-mas*) and lady-fern (*Athyrium filix-femina*). Many remnants of the Caledonian pine forest (Pinetum scoticae) still exist within the area on humus-iron, iron and humus podzols, notable examples being those at Glentanar and Ballochbuie on Deeside, Abernethy and Rothiemurchus on Speyside and the Black Wood of Rannoch (Steven and Carlisle, 1959).

Oak and birch woodland is widespread on the brown forest soils and humus-iron podzols of valleys and foothills. The woodlands of the east have generally been planted, although long enough established in many cases for the vegetation to have become 'natural' when grazing animals have been excluded. Two main associations have been distinguished (Birse, 1982). The first is eastern highland oakwood and birchwood (Trientali-Betuletum pendulae) which extends from the north down the eastern Grampians and as far west as Glen Affric. The oakwood at Dinnet on Deeside is a fine example. Southern oakwood (Galio saxatilis-

41

Quercetum) — the second association — occurs only as a remnant round the Moray Firth, particularly in the woodlands of Cawdor and Darnaway and only becomes important in Scotland to the south of Strathmore. The oak and birch of the west (Blechno-Quercetum) are remnants of the former natural woodland cover and occur extensively from Loch Lomondside to Inverness, although the overlap with the eastern association may vary considerably. Ash-oak wood (Primulo-Quercetum) has a distribution much like that of western oakwood and occupies similar sites on steep slopes but on better base status brown forest soils. The canopy may be either of ash (*Fraxinus excelsior*) with an understory of hazel (*Corylus avellana*) or of hazel alone. Hazelwood in the east is considered to be a part of elmwood (Querco-Ulmetum glabrae) which occupies steep-sided gullies and mixed bottom land on brown forest soils. The dens of Fife are notable sites of this broadleaved woodland association. Flush alderwood (the *Crepis paludosa–Alnus glutinosa* Association) is a local community of acid, peaty and humic gleys associated with springs on valley sides and river terraces.

Several fragmentary shrub communities of limited areal extent occur throughout Eastern Scotland, including wet birchwood (the *Sphagnum palustre–Betula pubescens* Community), common sallow scrub (the *Salix atro-cinerea* Community) and bog myrtle scrub (Myricetum galis) on gleys, peaty gleys and peat, and broom and gorse scrub (Sarothamnion), blackthorn scrub (the *Primula vulgaris–Prunus spinosa* Association) and boreal juniper scrub (Trien-tali-Juniperetum communis) on brown forest soils.

Foreshore and dunes

There are many considerable areas of windblown shelly and non-shelly sand round the east coast from Nigg in the north to Largo Bay in the south, the more extensive deposits being at Culbin Sands, Rattray Head, the Sands of Forvie, Buddon Ness and Tentsmuir. A typical sequence of dune communities colonizes the varying stages of dune development although every phase may not be present in any one locality. The tidal detritus of the foreshore is the site of orache strand-line (the *Salsola kali–Atriplex glabriuscula* Association) and this is succeeded by fore-dunes with northern sea couch-grass vegetation (Elymo-Agropyretum boreo-atlanticum). The sand of yellow or mobile dunes is bound together by the northern marram grass community (Elymo-Ammophiletum) and the grey or stable dunes and flat freely drained links to the rear of the dune system support a closed turf of milk vetch–red fescue dune pasture (Astragalo-Festucetum arenariae) on soils in which some profile development can be seen. The last-named community may be radically altered by heavy grazing and dunging to meadow-grass–bent pasture (the *Galium saxatile–Poa pratensis* Community) or modified for use as golf links. In some instances, a degree of arable cultivation is possible. Occasionally, a pioneer dune community of lyme-grass (Potentillo-Elymetum arenariae) which does not form an integral part of the sand-dune succession may be present above the drift line. The noncalcareous and calcareous gleys of dune slacks and wet depressions to the rear of the dune systems may carry a range of swamp, sedge and rush communities of which silverweed pasture (the *Potentilla anserina–Carex nigra* Community) is perhaps the most widespread. Because of their range of plant communities and unique species, many of these dune systems are conserved as sites of scientific interest, notable examples being Loch of Strathbeg on Rattray Head, St Cyrus near Montrose and Tentsmuir and Dumbarnie Links in Fife. Dumbarnie Links is one of the few stations of the variegated horsetail community (the *Anagallis tenella–Equisetum variegatum*

Association). A form of Atlantic heather moor (Carici binervis-Ericetum cinereae)—characterized by the presence of sand sedge (*Carex arenaria*) and birdsfoot trefoil (*Lotus corniculatus*)—occurs on the immature links soils at Delnies and the gravelly storm beaches at Findhorn.

Saltings

Saline alluvial soils occur around most of the estuaries and the two most commonly encountered communities of saltings are those of sea poa salt-marsh (Puccinellietum maritimae) and mud rush salt-marsh (Juncetum gerardii). The first association is found at, or slightly below, the high-water mark and the second at a slightly higher level. The salt marshes on the north shore of the Tay Estuary are the site of one of the largest continuous reed beds in Britain (Phragmitetum communis) and reed-cutting is still carried out commercially.

2   The Soil Map Units

The soils in Eastern Scotland are grouped into 233 map units which are numbered and defined nationally: units unlisted in the numbered sequence occur in other areas. Each map unit is recognized on the basis of one, or more than one, dominant major soil subgroup which, together with subsidiary subgroups, is related to a landform with specific rockiness and slope classes. Characteristic vegetation reflecting the relationship of climate and soil type is also listed. These soil map units are assigned to 56 soil associations according to parent materials based on lithology and drift type. In general, the units in each association are arranged in an ascending altitudinal sequence; the corresponding units derived from comparable drift types but different lithologies usually show marked similarities. With the exception of the Alluvial Soils and Organic Soils which are described first, the soil associations are arranged in alphabetical order.

Reference to Table A shows that Eastern Scotland is dominated by soil associations derived from acid parent materials. Almost half of the land area (45 per cent) is covered by four associations, namely the Arkaig, Corby, Countesswells and Strichen Associations. Peatlands and alluvial soils occupy another 12 per cent. The Balrownie, Foudland and Tarves Associations, virtually co-equal in area, occupy another 15 per cent; in other words almost three-quarters of the total land area is occupied by nine soil associations out of a grand total of 56 associations covering Eastern Scotland. Of the remaining 47 associations, 15 each occupy 0.1 per cent of the total area, or less.

THE ALLUVIAL SOILS

(Map units 1 and 2)

Alluvium comprises the products derived from the post-glacial erosion cycle and laid down as sediments from a suspension in water. Such parent materials may be deposited in riverine, lacustine, estuarine and marine conditions. They are normally well sorted with the modal grade size depending upon the current of the suspension; textures thus range from clay to gravel with variations occurring both horizontally and vertically. Profile development is absent or negligible except in the oldest deposits where an incipient B horizon often exists.

Under certain estuarine conditions, for example, during the immediate post-glacial period in the Rivers Tay and Forth, the brackish conditions resulting from tidal incurrents led to the widespread flocculation and deposition of silt and clay particles. Often more than 3 metres thick, these extensive stone-free deposits form the parent materials of the Stirling Association and are grouped with the closely related parent materials of the Duffus, Carbrook and Pow Associations; these were laid down respectively under lacustrine, high raised beach and estuarine conditions. Under more saline conditions in contemporary estuarine or in sheltered marine environments, a coarser and poorly drained alluvium with a sandy or loamy texture forms the saltings of map unit 2.

In riverine conditions, the suspended load reflects the nature and climate of the catchment areas whereas the deposition is determined largely by the interaction of the profile and gradient of the river with its velocity. Within Eastern Scotland, these water-courses range from steeply falling rivers and streams within narrow valleys in the highlands to major, meandering rivers of the lowlands. Often, the areas of alluvial soils are so small and occur in such a close association with fluvioglacial sands and gravels that they are included within the appropriate unit, usually *map unit 98* (the Corby Association); in Fife, however, the alluvium is frequently a component of *map unit 200* (the Eckford Association). Flash-flooding is a common hazard in the highland valleys and is associated with a constant cycle of erosion and deposition. Periodic flooding occurs in the haughlands of some of the major rivers though the widespread construction of embankments has reduced the threat of erosion to the surrounding unconsolidated alluvium; likewise, deposition is prevented by the levees under normal conditions.

Along the coast and below 10 metres, especially around the Moray Firth, immature soils dominate the raised beaches formed from marine alluvium. Such soils often include peaty alluvium and poorly drained alluvium at the landward edge of each platform. In places, bands of silt and clay occur, indicating the lagoonal nature of the deposit during the emergence of the raised beaches. Most of these soils are included within the Nigg/Preston Associations.

Lacustrine alluvium is common across the lowlands and is associated occasionally with lochs in the highlands, for example, Loch Morlich and Loch Moy; most deposits, however, fully occupy post-glacial loch sites.

Excepting the above associations, all other major areas of alluvial soils have been grouped together within map unit 1, irrespective of texture and drainage. Alluvial parent materials, regardless of their depositional environment, cover approximately 1000 square kilometres in Eastern Scotland.

Map unit 1 Extending over 674 square kilometres (2.8 per cent of the land area), the unit comprises riverine and lacustrine alluvium. Both mineral and peaty soils are included, the peaty alluvial soils being more common in the western highland valleys. Normally poorly or very poorly drained, these latter soils include redistributed peat, interbedded organic and mineral soils and mineral alluvial soils with a surface peat formation. The unit occurs over a wide altitudinal range, from sea level to approximately 500 metres in Glen Derry, within the climate regions of warm moderately dry lowlands to cold wet uplands. Most of the larger areas, however, lie below 200 metres in the warm moderately dry lowlands and are associated with the haughlands of the major rivers. In some rivers, for example, the River Spey, the lower reaches are strongly braided, the combination of the braiding and the shingle bars providing constant flooding problems at the mouth. Severe flooding is also a hazard in the lower reaches of the faster flowing rivers, for example, the River Beauly and the River Findhorn (11 metres fall per kilometre), where levees have been breached occasionally and large quantities of sand and gravel deposited across the adjacent alluvial terraces. Deltaic alluvium also is commonly associated with flooding in certain river valleys; it forms where very fast flowing tributaries enter the main channel. The annual flooding in the middle reaches of the River Spey is, in part, the consequence of alluvial cones of sand and gravel formed at the mouths of the Rivers Nethy, Druie and Feshie. It is also likely that the Feshie delta has facilitated the formation of the expanse of peaty alluvium above Loch Insch. In Glen Truim, the fans emerging from the deeply incised valleys in the hills above Drumochter Lodge have coalesced to form a small piedmont alluvial plain. Occasionally such fans, with their immature soils, are incised into terrace or outwash sands and gravels which are characterized by well-developed podzols, for example the Allt a Mharcaith east of Feshiebridge.

Sandy loams, frequently stone-free and more than 0.5 metres thick, dominate most of the alluvial terraces of the lowland rivers. These soils are inherently fertile and free-draining, overlying loamy sands or sands which are underlain by gravel at approximately 1 metre from the surface. Similar patterns are repeated in other parts of the major rivers wherever the gradient and flood plain permit the deposition of these finer sediments, for example, above Tomatin on the River Findhorn. The terraces are often more than 1 kilometre wide, reaching their maximum width of some 3 kilometres near Caputh on the River Tay. Sometimes a slightly higher terrace exists above the flood plain with profile development differentiating a weakly expressed, yellowish B horizon.

Occasionally, the meandering of the major rivers in wide flood plains has led to the formation of ox-bow lakes. Most have been subsequently infilled by fine sediments usually with a silty upper horizon capped by a peaty topsoil; they are invariably poorly drained and situated near the outer margins of the flood plain. Mottling, especially in the form of iron tubes surrounding roots, is common in the upper profile, contrasting sharply with the blue-grey colours associated with the permanent, high water-table. In other areas, where the drainage is imperfect, iron mottles are widespread and manganese staining or pans often occur at textural interfaces.

Lacustrine alluvium is more locally derived and finer textured than riverine material, normally stone-free and ranging from sandy loam to silty clay. Sometimes containing buried topsoil or peaty horizons, most are distinctly banded and poorly drained. In the lowlands, they often have marl beds; these layers of molluscan shells were frequently exploited in the nineteenth century as a source of lime. Today, lacustrine sites are occasionally excavated as fish ponds.

In conjunction with the favourable lowland climate, the downstream alluvial terraces of the major rivers, which drain Tayside and the Moray Firth catchment, contain some of the most fertile soils in Eastern Scotland; the soils are capable of producing high yields of a wide range of crops. With increasing altitude and more adverse climate factors, especially rainfall, these soils have progressively reduced cropping flexibility and yields; in the wetter western areas, the unit is largely restricted to semi-permanent grassland. Because of their inherent poor drainage, many of the lacustrine alluvial areas are restricted to grass reclamation or are maintained as rough grazings. The grazings often comprise rush pastures with moderate values though, where the sites are developed from peaty alluvium and support bog heather moors (Narthecio-Ericetum tetralicis) or blanket bogs (Erico-Sphagnetum papillosi), the grazing values are low. In the highland valleys, the alluvial soils usually form an integral part of the agricultural system; they are often the only areas capable of producing winter feed for a grazing population which would otherwise be impossible to sustain.

In one estate, an unusual use was made occasionally of tiny alluvial flats in very high-level catchments; there, oats were grown and stooked to supplement the diet of the local grouse population.

Lack of fall in the dominantly level sites and poor soil structure are major limitations in the drainage of poorly drained alluvium. Because of the uniform particle grading and the high water-table, the soils are often thixotropic. Revetting with timber is a necessary and costly procedure in trunk drainage schemes where low-angle slopes are not possible. A recent development, still awaiting evaluation, has been the novel introduction of a pump scheme allied to major sumps excavated parallel to the Spynie canal in the poorly drained alluvium north of Elgin.

Map unit 2   This map unit is one of the smaller units in Eastern Scotland (Sheet 5) and occupies only 3 square kilometres, less than 1 per cent of the association. It consists of salt marshes and has been mapped in two areas between Nairn and Findhorn, the Buckie Loch area around the Culbin foreshore and the landward area of the Nairn bar. Other minor areas of saltings too small to identify individually at the 1:250 000 scale, for example, around Nigg Bay, and the Carse of Delnies, have been included within the associated *map units 382* or *421*.

The landform is restricted to the lowest raised beach around the high-water mark of ordinary spring tides. Slopes are negligible and non-rocky. 'The action of the tides is probably more important than the moisture balance of the climate and the extreme maritime position of the soils outweighs the influence of oceanity' (Birse *et al.*, 1976). Average annual rainfall of below 600 millimetres is one of the lowest in Scotland and the areas occur in the warm, moderately dry climate region.

The soils are exclusively poorly or very poorly drained saline alluvial soils. Derived from marine sands, they are immature and constantly subjected to erosion and accretion, especially accretion by aeolian sand. Depending upon their position relative to the tides, the degrees of flooding and salinity vary widely. The former ranges from a twice daily inundation to flooding only at high spring tides. Extremely high values for exchangeable cations in the surface horizons reflect the dominating tidal influence. Sodium replaces calcium as the dominant cation and values of 60 milli-equivalents per 100 grammes have been recorded. A permanent water-table is encountered usually below 40 centimetres where the intensity of the anaerobic reducing conditions is indicated by the greenish grey colours. The pH at the surface is near neutral but, in the absence of shell fragments, rapidly becomes acidic, and pH values frequently drop to 3.5–4.5 in the lower horizons.

Land use is restricted to rough grazing either by sheep or cattle. The plant communities are specific and characteristic. At the ordinary high tide mark the sea poa salt-marsh (Puccinellietum maritimae) is associated with the diurnal flooding whereas at slightly higher levels it is replaced by the mud rush salt-marsh community (Juncetum gerardii). The latter extends above the reach of high spring tides but is still influenced by sea-water throughout the year.

THE ORGANIC SOILS

(Map units 3 and 4)

Organic Soils comprise basin, valley and blanket peats, and cover 2062 square kilometres (8.4 per cent of the land area) in Eastern Scotland. They occur from below the tide-mark in Burghead Bay to over 1060 metres in the Cairngorm Mountains and form the fourth largest association in the region. Peat also forms an integral part of soil complexes in the foothills, uplands and mountains where it is a major component of 40 map units and a minor component in another 39 units. Occasionally, the relationship is distinctive, as for example in the widespread hummocky moraine fields and the fluvioglacial outwash plains. The true areal extent of the peatlands is therefore much greater than that of the Organic Soils.

Conventionally, peat is regarded as a soil having more than 60 per cent organic matter. Although there is no exact line of demarcation, the Soil Survey of Scotland originally defined a thickness of 12 inches (30 centimetres) as necessary to separate the peatlands. This criterion was used until 1972 by which time experience indicated that a greater logicality in separation was obtained by using a thickness of 50 centimetres; the first 1:63 360 map to show this division was Latheron and Wick (Sheets 110, 116 and part of 117) (Futty and Dry, 1972). Since that date, apart from the unusual use of 45 centimetres in one instance, most maps and publications, including the present series, have adopted 50 centimetres as standard.

Despite the work of many, amongst them Robertson (1933), Fraser (1954) and Hulme (1980), a universally accepted classification for peatlands in Scotland has still to emerge. Fraser (1954) introduced the most comprehensive scheme, dividing peat into climate (or zonal) bogs and intrazonal bogs; he equated the former with blanket bog and subdivided it into (A) blanket bogs of the western and northern oceanic regions, (B) peat bogs of hills developed under high rainfall and low temperatures and (C) arctic-alpine-climate bogs of some alpine plateaux, for example, the Cairngorm Mountains. The intrazonal bogs are divided into peats developing in or on free water and those developing on water-logged or intermittently flooded mineral soil and vegetation. These groups are further subdivided. The Soil Survey of Scotland, however, maintained a simplistic division of peatlands into hill peat and basin peat from 1954, when its first 1:63 360 soil map was published, until 1972. Then Futty and Dry (1972) and Ragg et al. (1975) included valley peats with the basin peats but did not distinguish them cartographically. Thereafter, Walker et al. (1976) recognized and mapped a triple division of (A) basin peat, (B) valley and terrace peats and (C) blanket peat (hill peat), the second element being equated with the valley bogs and impeded stagnant bogs identified by Fraser. In an attempt to develop a broad and generally applicable classification system, Hulme (1980), proposed a triple topographic division of confined mire, partly confined mire and unconfined mire. His classes have obvious affinities with those of earlier systems. During the present survey, only two categories, have been recognized, (A) basin and valley peats, including terrace peats, (map unit 3) and (B) blanket peat (map unit 4).

Within Eastern Scotland, the basin peats occur throughout the lowlands in confined depressions. The valley and terrace peats, which are intrazonal and partly confined, are more characteristic of the foothills and uplands. They occur on level or gently sloping sites; apart from terraces and valley floors, these include bench-like embayments in the major valleys and various planation surfaces. Such sites are formed mainly in thick deposits of till. Occasionally, these partly confined peats are adjacent to extensive areas of hill peat and have been included within the blanket peat (*map unit 4*). Confined peats, or basin peats, within upland and mountain regions are blanketed usually by the development of surrounding ombrogenous hill peat and are similarly mapped within *map unit 4*.

Map unit 3 Covering 272 square kilometres, the unit consists of basin and valley peats, including terrace peats. The true basin or confined peats, developing from a lacustrine origin, are restricted mainly to the North-East

Lowlands and the upper Forth valley. Minor areas occur in Fife, especially around Loch Glow, and on the raised beaches of the Moray Firth Lowlands; throughout Strathmore and the Howe of the Mearns, no peatland is large enough to reproduce at the 1:250 000 scale.

The north-eastern peats occur mostly between Peterhead and New Pitsligo where they are underlain by gleys of the Peterhead and Tarves Association. Sandy clay loam to clay textures, coupled with the massive structure of these soils, effectively impede internal drainage and lead to the waterlogged conditions necessary for peat formation. Most of these peats are remnants of much wider peatlands as is the largest basin peat within Eastern Scotland, Flanders Moss in the upper Forth valley; this peat once extended downstream to Stirling and is underlain by the massive clays of the Stirling Association. Thickness of all these peats varies enormously from 1 to 8 metres with an average around 5 metres. Area also varies widely and ranges from 50 to 600 hectares except at Flanders Moss which is approximately 2000 hectares.

Climate regions include mainly the warm and fairly warm dry lowlands of the north-east, with 700 to 900 millimetres average annual rainfall, and the warm wet lowlands of the upper Forth valley with 1200 to 1600 millimetres rainfall.

Outwith the lowlands, the valley and terrace peats have a wide altitudinal range because of their azonal nature, for example from 300 metres east of Nethybridge to nearly 750 metres around the Edendon Water north of Glen Garry. Usually fairly extensive, these peatlands cover up to 1100 hectares and occur in the climate regions of cool and cold wet foothills and uplands. Rainfall is mostly between 1000 and 1400 millimetres and reaches a maximum of 1800 millimetres to the south-west of Loch Rannoch and around the headwaters of the River Spey. These peatlands overlie a wide variety of parent materials and soil profiles though in all cases an impeding horizon has induced surface waterlogging and the subsequent formation of peat. Induration and iron pans are amongst the commonest causes.

Apart from the traditional use of lowland bogs for fuel, there is an increasing commercial exploitation of such deposits; specialized machinery either extrudes sod peats or the surface is milled prior to the production of briquettes. The production of peat for horticultural purposes is also expanding rapidly. Other peatlands are used as rough grazing and in many instances have been reclaimed successfully to grassland, especially in the low rainfall areas. Afforestation, mainly by lodgepole pine but occasionally by Sitka spruce, was steadily increasing in the fifties and sixties before industry recognized the extraction value of these deposits.

Map unit 4   Blanket peat is the extreme form of climatic peat which reaches its maximum European development in the British Isles. In Eastern Scotland, this ombrogenous, unconfined peatland covers 1790 square kilometres (7.2 per cent of the land area) and, with minor exceptions, is correlated directly with hill peat. It is extensive throughout the North-East Grampian Highlands and dominates the Monadhliath Mountains, extending northwards to some of the higher summits within the Moray and Banff foothills. In the South-West Grampian Highlands, however, the more strongly dissected and steeper sloping topography militate against the widespread development of peat; the main areas are the source catchments of the Rivers Almond, Quaich and Lednock south of Loch Tay and the upper hill slopes south-east of Glen Artney. Within the Central Lowlands the unit is restricted to the western summits of the Ochil Hills and the northern sector of the Gargunnock Hills.

Most of the major blanket peats occur in the cool and cold wet uplands with lesser areas in the very cold and wet mountain climate region, for example, the Drumochter Hills; the average annual rainfall ranges from 1000 to 2000 millimetres, excepting the South-West Grampian Highlands where the high ground has between 2000 and 3200 millimetres rainfall. Generally occurring from 500 to 800 metres, the peat envelops most slopes below 10 degrees to a depth of 2 or 3 metres; in cols and local basins, the thickness may exceed 3 or 4 metres. Above the subalpine soil zone, the map unit is confined to prominent planation levels or plateau summits. Although the development of thick blanket peat is restricted by slopes much above 15 degrees, by highly permeable substrata and by soil parent material with a high base status, these limitations are gradually subordinated as the effective climate wetness becomes paramount. Thus, towards the south-west, shallow blanket peat, less than 1 metre thick, can mantle slopes up to 20 degrees and possibly beyond.

Gully erosion is widespread and the eroded phase of map unit 4 covers 821 square kilometres, almost half the total area of blanket peat in Eastern Scotland. Radial in distribution on the steeper slopes, which surround smooth, rounded summits and characteristically dendritic in flatter sites and cols, the erosion has often penetrated the underlying mineral soil. Sometimes the dessication, following this dissection of the mantle, eventually leads to sheet erosion by wind and water, albeit on a minor scale to date with single areas seldom exceeding 1 hectare.

On many severely exposed summits especially, around 800 metres where the peaks barely emerge above the blanket peat, for example, in the Monadhliath Mountains and the Ladder Hills, there are isolated tabular areas of peat usually about 1 metre thick. They are seldom more than 10 or 20 metres wide and are completely surrounded by relatively large expanses of alpine or subalpine soils. Almost certainly they represent the remnants of a much more extensive cover which probably blanketed every gently sloping summit at these altitudes and above, except those which were extremely exposed.

In general, humification increases towards the base of a peat deposit, the nature of which is determined largely by the surface vegetation during development and the height of the water-table. On the drier sites, the greater degree of oxidation leads to a more highly humified or amorphous peat, as is found in most shallow blanket peats in Eastern Scotland. Deeper peats, though composed mainly of *Sphagnum* and *Eriophorum* spp., show a distinct stratification resulting from different degrees of humification; most horizons are slightly or moderately humified. The stratification is occasionally enhanced by the presence of a basal layer of macroscopic birch remnants and by a layer of pine stumps in a horizon developed largely from *Calluna vulgaris*. Lewis (1911) also identified a layer of arctic and subarctic vegetation, mainly willows, below the birch horizon.

Plant communities below the subalpine soil zone comprise various forms of blanket bogs (Erico-Sphagnetum papillosi). An eastern and a western grouping can be recognized with the western blanket bogs being dominated by flying bent–bog, myrtle bog, flying bent–cotton-grass bog and common cotton-grass bog. The eastern peats, mainly associated with the drier uplands and plateaux, consist of northern blanket bog and upland blanket bog. On the higher levels, where erosion and subsequent drying out of the peat mantle is common, the *Sphagnum* species are replaced by *Racomitrium lanuginosum* and dwarf shrubs, especially *Empetrum nigrum* and *Vaccinium uliginosum*. The *Racomitrium* is found along with lichens of the *Cladonia sylvatica* group in the undisturbed

Table A Areas of soil map units

ASSOCIATION (sq. km., % Land Area)	MAP UNIT	AREA (sq. km)	% Land Area	% Association
ALLUVIAL SOILS (677 sq. km., 2.8%)	1	674	2.8	100
	2	3	<0.1	<1
ORGANIC SOILS (2062 sq. km., 8.4%)	3	260	1.1	13
	3e	12	<0.1	<1
	4	969	4.0	47
	4e	821	3.3	40
ABERLOUR (271 sq. km., 1.1%)	5	46	0.2	17
	6	60	0.2	22
	7	33	0.1	12
	9	24	0.1	9
	10	31	0.1	11
	11	20	<0.1	7
	12	24	<0.1	9
	13	5	<0.1	2
	14	9	<0.1	3
	15	19	<0.1	7
ARDVANIE	17	29	0.1	100
ARKAIG (3275 sq. km., 13.3%)	19	90	0.4	3
	20	254	1.0	8
	21	92	0.4	3
	22	413	1.7	13
	23	332	1.4	10
	25	62	0.3	2
	26	236	1.0	7
	27	52	0.2	2
	28	368	1.5	11
	29	196	0.8	6
	30	220	0.9	7
	31	52	0.2	2
	32	13	<0.1	<1
	33	377	1.5	12
	34	200	0.8	6
	35	140	0.6	4
	36	178	0.7	5
BALROWNIE (1239 sq. km., 5.0%)	41	948	3.9	76
	42	97	0.4	8
	43	82	0.3	7
	44	57	0.2	5
	45	26	0.1	2
	46	29	0.1	2

ASSOCIATION (sq. km., % Land Area)	MAP UNIT	AREA (sq. km)	% Land Area	% Association
BERRIEDALE	61	<1	<0.1	100
BOGTOWN	70	10	<0.1	100
BRAEMORE/ KINSTEARY (30 sq. km., 0.1%)	71	7	<0.1	25
	72	1	<0.1	<5
	73	8	<0.1	25
	74	14	<0.1	45
BRIGHTMONY	76	21	<0.1	100
CARPOW/PANBRIDE	89	129	0.5	100
CORBY/BOYNDIE DINNET (1869 sq. km., 7.6%)	96	14	<0.1	<1
	97	943	3.8	51
	98	471	1.9	25
	100	214	0.9	12
	101	229	0.9	12
CORRIEBRECK	107	<1	<0.1	100
COUNTESSWELLS/ DALBEATTIE/ PRIESTLAW (2253 sq. km., 9.2%)	115	820	3.3	36
	116	121	0.5	5
	117	292	1.2	13
	118	87	0.4	4
	119	24	<0.1	1
	122	<1	<0.1	<1
	123	125	0.5	6
	125	4	<0.1	<1
	126	58	0.2	3
	127	57	0.2	3
	128	3	<0.1	<1
	129	51	0.2	2
	131	9	<0.1	<1
	134	70	0.3	3
	135	153	0.6	7
	136	311	0.9	9
	137	168	0.7	8
CRAIGELLACHIE/ POLFADEN	140	12	<0.1	100
CROMARTY/ KINDEACE (174 sq. km., 0.7%)	144	40	0.2	25
	145	115	0.5	65
	146	19	<0.1	10

Table A Areas of soil map units

ASSOCIATION (sq. km., % Land Area)	MAP UNIT	AREA (sq. km)	% Land Area	% Association
DARLEITH/ KIRKTONMOOR (191 sq. km., 0.8%)	147	138	0.6	70
	149	<1	<0.1	<5
	150	18	<0.1	10
	153	10	<0.1	5
	154	8	<0.1	<5
	158	17	<0.1	10
DARVEL	163	36	0.2	100
DEECASTLE	165	33	0.1	100
DOUNE	168	37	0.2	100
DREGHORN	169	48	0.2	100
DULSIE (189 sq. km., 0.8%)	172	79	0.3	40
	173	16	<0.1	10
	174	4	<0.1	<5
	175	90	0.4	50
DURNHILL (297 sq. km., 1.2%)	181	8	<0.1	3
	182	134	0.6	45
	183	4	<0.1	1
	185	1	<0.1	<1
	187	2	<0.1	<1
	189	4	<0.1	1
	192	26	0.1	9
	193	40	0.2	14
	194	10	<0.1	3
	195	68	0.3	23
ECKFORD/ INNERWICK (97 sq. km., 0.4%)	199	3	<0.1	<5
	200	94	0.4	95
ELGIN (62 sq. km., 0.3%)	201	4	<0.1	5
	202	57	0.2	90
	203	1	<0.1	<5
ETHIE	204	18	<0.1	100
FORFAR (375 sq. km., 1.5%)	237	66	0.3	18
	238	7	<0.1	2
	239	302	1.2	80

ASSOCIATION (sq. km., % Land Area)	MAP UNIT	AREA (sq. km)	% Land Area	% Association
	240	18	<0.1	1
	241	103	0.4	8
	243	878	3.6	67
	244	73	0.3	6
	245	121	0.5	9
	246	9	<0.1	<1
	248	5	<0.1	<1
FOUDLAND (1310 sq. km., 5.3%)	250	7	<0.1	<1
	252	17	<0.1	1
	253	19	<0.1	2
	255	19	<0.1	2
	256	24	<0.1	2
	257	13	<0.1	<1
	258	4	<0.1	<1
FRASERBURGH	259	6	<0.1	100
GLENEAGLES/ AUCHENBLAE	273	144	0.6	100
GOURDIE/CALLANDER/ STRATHFINELLA (350 sq. km., 1.4%)	274	202	0.8	58
	275	63	0.3	18
	276	85	0.3	24
	277	<1	<0.1	<1
HATTON/TOMINTOUL/ KESSOCK (232 sq. km., 1.0%)	281	51	0.2	22
	282	89	0.4	38
	283	7	<0.1	3
	284	29	0.1	13
	285	27	0.1	12
	286	29	0.1	13
HINDSWARD	291	27	0.1	100
INSCH (411 sq. km., 1.7%)	316	257	1.1	63
	317	55	0.2	13
	318	76	0.3	19
	319	1	<0.1	<1
	320	6	<0.1	2
	323	7	<0.1	2
	324	1	<0.1	<1
	326	1	<0.1	<1
	328	2	<0.1	<1
	329	2	<0.1	<1
	330	3	<0.1	<1

Table A Areas of soil map units

ASSOCIATION (sq. km., % Land Area)	MAP UNIT	AREA (sq. km)	% Land Area	% Association	ASSOCIATION (sq. km., % Land Area)	MAP UNIT	AREA (sq. km)	% Land Area	% Association
KIPPEN/LARGS (95 sq. km., 0.4%)	337	50	0.2	55	SKELMUIR (20 sq. km., ‹0.1%)	462	11	‹0.1	55
	338	7	‹0.1	10		463	9	‹0.1	45
	339	22	‹0.1	25	SORN/HUMBIE/ BIEL (5 sq. km., ‹0.1%)	466	1	‹0.1	20
	341	7	‹0.1	10		467	4	‹0.1	80
	343	3	‹0.1	‹5	SOURHOPE (545 sq. km., 2.2%)	472	405	1.7	74
	344	6	‹0.1	5		473	4	‹0.1	‹1
LAURENCEKIRK	368	91	0.4	100		474	12	‹0.1	2
LESLIE (29 sq. km., 0.1%)	369	11	‹0.1	40		475	8	‹0.1	2
	370	15	‹0.1	50		476	72	0.3	13
	371	3	‹0.1	10		479	44	0.2	8
LINKS (142 sq. km., 0.6%)	380	126	0.5	90	STIRLING/DUFFUS/ POW/CARBROOK (240 sq. km., 1.0%)	487	24	‹0.1	10
	382	16	‹0.1	10		488	216	0.9	90
MILLBUIE (199 sq. km., 0.8%)	405	12	‹0.1	5	STONEHAVEN (253 sq. km., 1.0%)	490	180	0.7	71
	406	187	0.8	95		491	10	‹0.1	4
MOUNTBOY	414	214	0.9	100		429	26	0.1	10
NIGG/PRESTON (119 sq. km., 0.5%)	420	69	0.3	60		493	29	0.1	12
	421	50	0.2	40		494	3	‹0.1	1
NOCHTY	422	10	‹0.1	100		495	3	‹0.1	1
NORTH MORMOND/ ORTON (233 sq. km., 0.9 %)	423	21	‹0.1	9		496	2	‹0.1	‹1
	424	34	0.1	15	STRICHEN (3827 sq. km., 15.6%)	497	430	1.8	11
	425	173	0.7	74		498	938	3.8	25
	426	5	‹0.1	2		499	453	1.8	12
ORDLEY/ CUMINESTOWN (166 sq. km., 0.7%)	427	12	‹0.1	5		500	182	0.7	5
	428	154	0.6	95		501	88	0.4	2
PETERHEAD (138 sq. km., 0.6%)	429	71	0.3	50		502	24	‹0.1	‹1
	430	67	0.3	50		503	347	1.4	9
ROWANHILL/GIFFNOCK/ WINTON (600 sq. km., 2.4%)	444	92	0.4	15		504	112	0.4	3
	445	444	1.8	74		505	237	1.0	6
	446	49	0.2	8		506	418	1.7	11
	447	5	‹0.1	‹1		507	151	0.6	4
	448	10	‹0.1	2		508	10	‹0.1	‹1
SABHAIL/ MOUNTEAGLE (129 sq. km., 0.5%)	454	105	0.4	85		509	61	0.3	2
	457	6	‹0.1	‹5		512	144	0.6	4
	457	18	‹0.1	15		513	169	0.7	4
						514	54	0.2	1
						515	9	‹0.1	‹1

53

Table A Areas of soil map units

ASSOCIATION (sq. km., % Land Area)	MAP UNIT	AREA (sq. km.)	% Land Area	% Association	ASSOCIATION (sq. km., % Land Area)	MAP UNIT	AREA (sq. km.)	% Land Area	% Association
	517	629	2.6	50	TIPPERTY	545	51	0.2	100
	518	262	1.1	21					
	520	104	0.4	8		565	1	‹0.1	‹5
	521	79	0.3	6	TYNET (44 sq. km., 0.2%)	566	30	0.1	70
	522	60	0.3	5		567	13	‹0.1	30
	523	13	‹0.1	1					
TARVES (1255 sq. km., 5.1%)	524	5	‹0.1	‹1	ROCK		19	‹0.1	
	525	13	‹0.1	‹1					
	527	11	‹0.1	‹1	BUILT-UP AREAS		208	0.9	
	529	3	‹0.1	‹1					
	530	9	‹0.1	‹1					
	532	28	0.1	2					
	533	34	0.1	3					
	534	5	‹0.1	‹1					

Areas in this table have been estimated by point-count methods. Care should be exercised in calculations involving units of less than 10 square kilometres. Discussion of method and estimation of error is contained in Handbook 8.

blanket peats. In the subalpine and alpine soil zones, the peat is characterized by *Rubus chamaemorus* (Rhytidiadelpho-Sphagnetum fusci).

Land use is almost entirely devoted to rough grazing, mostly of low value. Within the eroded phase, the intense hagging effectively downgrades much of this grazing. Unlike many of the blanket peat areas north of the Great Glen, especially in Caithness, the hill peat is too high for commercial forestry.

THE ABERLOUR ASSOCIATION

(Map units 5–7, 9, 10–15)

The Aberlour Association is derived from a mixture of granite and acid schists or from granulites intruded by multiple granite veining, including granulites with variable intrusions of felsites. These rock types have produced a gritty sandy loam, locally loam, drift, which is stony and frequently very stony.

Some 271 square kilometres, a little over 1 per cent of the land area, have been mapped as the Aberlour Association. It is, however, only a locally extensive association. Major locations include the valleys in the north western and southern Monadhliath Mountains, the Nairn plateau, Rannoch and Gaick Forests and lower Strathspey.

The soils have a fairly wide altitudinal distribution from 200 to 900 metres and an average annual rainfall of 750 to 2200 millimetres. Humus-iron podzols, peaty podzols, peaty gleys and peat are the dominant component soils. In addition, there are locally extensive areas of noncalcareous gleys, brown forest soils, sub-alpine and alpine soils, peaty rankers and lithosols.

Because of the ranges of altitude and rainfall, the land capability for agriculture is varied. Although some of the land has a moderate capability for the production of a narrow range of crops, limitations such as climate, wetness, topography and soil restrict most of this association to moderate or poor grassland. On the exposed upper hill slopes, land use is usually limited to rough grazings which are of relatively low grazing value.

Map unit 5 This unit consists of noncalcareous and peaty gleys with some humic gleys and peat, and covers an area of 46 square kilometres (17 per cent of the association). Confined to valley sides with concave, regular, gentle and strong, non-rocky slopes, the map unit has a narrow altitude range (200 to 350 metres) and an average annual rainfall between 750 and 1200 millimetres. Climatically, the unit is associated with fairly warm, moderately dry lowlands and with cool wet foothills and uplands.

Principal locations are the catchment of the Findhorn valley near Tomatin and Ferness, the areas around Aberlour in Speyside and around Errogie east of Loch Ness. Because of the lowland situation, the drift is often thicker than that of most other Aberlour map units and the texture is finer, usually a loam or sandy clay loam. Areas occur, however, on strong slopes where the drift is shallow and coarser, mainly a stony, gritty sandy loam, with rock close to the surface.

On the cultivated land, which occupies most of the map unit, permanent and ley pastures (Lolio-Cynosuretum) and arable crops are found; soft rush pasture (the *Ranunculus repens–Juncus effusus* Community) and sedge mires occur on the uncultivated gleys; the peaty gleys and peat have respectively bog heather moors (Narthecio-Ericetum tetralicis) and blanket bogs (Erico-Sphagnetum papillosi) as the main plant communities.

Map unit 6 Freely and imperfectly drained humus-iron podzols are the dominant soils, some of which are cultivated. There are minor areas of brown forest soils and gleys.

Map unit 6 occupies 60 square kilometres or 22 per cent of the association and is the most extensive unit of the Aberlour Association. Large areas of it are found in the triangle in Speyside formed by Rothes, Dufftown and Ballindalloch. The unit has the same altitude and rainfall range as *map unit 5*, 200 to 350 metres and 750 to 1200 millimetres respectively. It is associated similarly with the climate regions of fairly warm, moderately dry lowlands and with cool, wet foothills and uplands.

The landform is one of hills and valley sides with gentle and strong non-rocky slopes. The drift may have a stony, gritty, sandy loam texture where the rock is close to the surface on the hills and valley sides, but on the lower, gentle slopes a thick, stony gritty loam till is usually present.

The dominant plant community is dry boreal heather moor (part of Vaccinio-Ericetum cinereae) and is mainly associated with humus-iron podzols. Where the soils have been cultivated, there are permanent and ley pastures (Lolio-Cynosuretum).

Land use is generally moderate arable and mixed farming with long ley pastures. Where more severe climatic restrictions prevail, cultivation is difficult and land use restricted to permanent grassland.

Map unit 7 The unit is located mainly in the mid catchment of the River Findhorn and the upper catchment of the River Gairn; it covers some 33 square kilometres or 12 per cent of the association. Component soils are peaty podzols and peat with some peaty gleys and humus-iron podzols.

With an average annual rainfall of 900 to 1200 millimetres and an altitude range of 300 to 450 metres, map unit 7 is located in the cool, wet foothills and uplands and the cold, wet upland climate regions.

The landforms are hills and valley sides with strong and steep, non-rocky slopes. On strong, concave slopes in some locations, a thick till of a stony loam texture may be found, but normally the drift is a shallow, stony, gritty sandy loam.

Major limitations to agriculture include poor to very poor drainage, a thick organic horizon, climate and topography. Some areas have a substantial capability for grassland reclamation, but generally the land use is restricted to rough grazings.

The plant community present on the peaty podzols is moist boreal heather moor (part of Vaccinio-Ericetum cinereae). On the peaty gleys, which usually have thicker organic horizons, are bog heather moors (Narthecio-Ericetum tetralicis). Lowland and upland blanket bogs (parts of Erico-Sphagnetum papillosi) dominate the peat.

Map unit 9 Occupying 24 square kilometres or 9 per cent of the association, it is local in distribution and occurs in two areas to the south-west of Loch Ericht. Peaty podzols, peat and peaty gleys are the dominant component soils. The altitude range is wide, 300 to 650 metres, and the average annual rainfall is of the order 1600 to 2000 millimetres. Climatically, the map unit is associated with the cool, wet foothills and uplands.

The landform is one of hummocky moraine with strong to steep, non-rocky or slightly rocky, complex slopes. Soil textures tend to be very stony, gritty, loamy sands and the parent material is unsorted.

Poor or very poor drainage, a thick organic surface horizon and climate are all limitations to land use, but the most severe restriction is the hummocky topography. As a consequence, most of the area is restricted to rough grazings; grazing values are mainly low.

Moist boreal heather moor (part of Vaccinio-Ericetum cinereae) dominates the vegetation associated with the peaty podzols whereas bog heather moors (Narthecio-Ericetum tetralicis) and blanket bogs, (Erico-Sphagnetum papillosi) occur respectively on the peaty gleys and peat. On very minor local areas, grazing has produced a heath rush–fescue grassland (Junco squarrosi-Festucetum tenuifoliae).

Map unit 10 This map unit covers an area of 31 square kilometres (11 per cent of the association) in the northern hills, in particular the uplands of Morayshire and Nairnshire. Podzolic soils dominate, principally peaty podzols and humus-iron podzols, with peat in minor depressional sites. The landform is one of hills and valley sides with strong and steep moderately rocky slopes.

The altitude range is wide, 300 to 600 metres, and the average annual rainfall is of the order 750 to 1200 millimetres. Climatically, the map unit is associated with the cool, wet foothills and uplands or cold, wet uplands.

Apart from very local deposits of lodgement till on rock ledges, the soil parent material is a stony, gritty, sandy loam drift with a varying proportion of locally derived fragmented rock. On the steepest slopes, some colluviation has taken place forming a slightly finer-textured parent material.

The rockiness, organic surface horizons and severe slope complexity restrict the agriculture of this unit to rough grazing; the grazings are mainly of low value. It is, however, well suited to timber production and many areas have been afforested.

Where the organic surface horizon is not too thick and humus-iron podzols dominate, some limited reclamation to grassland may be possible amongst minor areas of outcropping rock.

Plant communities on the peaty podzols and humus-iron podzols are moist and dry boreal heather moors (parts of Vaccinio-Ericetum cinereae) and white bent–tussock-grass grassland (the *Cirsium palustre–Nardus stricta* Community).

Map unit 11   This unit may be regarded as similar to *map unit 10* but wetter. It covers only 20 square kilometres or 7 per cent of the association. Wet soils such as peaty gleys and peat dominate the unit although peaty podzols and peaty rankers are associated with the rocky sections. The altitude and average annual rainfall ranges are wide, 300 to 800 metres, and 1000 to 16000 millimetres respectively. Climatically, the map unit is located in the regions of the cool, wet foothills or cold, wet uplands. The largest area is west of Loch Caoldair, Strathmashie.

The associated landform is one of undulating hills with gentle and strong, moderately rocky slopes. Landform and parent materials are similar to those of *map unit 10*. Though lodgement till occurs in very local situations, a stony, gritty, sandy loam drift is widespread. A finer-textured colluvium has also been included within the map unit.

Limitations to land use are very severe; wetness, organic surface horizons, complex slopes and rockiness restrict the unit to rough grazing; grazing values are moderate around Loch Caoldair but low elsewhere.

Moist boreal heather moor (part of Vaccinio-Ericetum cinereae) is common on the peaty podzols, with lowland and upland blanket bogs (parts of Erico-Sphagnetum papillosi) and bog heather moors (Narthecio-Ericetum tetralicis) dominating respectively the peat and peaty gleys.

Map unit 12   This map unit occupies only 24 square kilometres or 9 per cent of the association and occurs on the complex eastern slopes of Carn Dearg in Rannoch Forest, the northern slopes of Meall Criaidh, Loch Ericht, in Gaick Forest and on the slopes north of Loch Earn.

With an average annual rainfall range of 1500 to 2300 millimetres at an altitude of 600 to 900 metres, the unit is associated with the climate regions of the cold, wet uplands and the very cold, wet mountains.

The soils are subalpine soils with some peat and rankers. Mountains with gentle to very steep, non- to very rocky slopes form the major landforms.

Locally derived shattered rock contributes substantially to the stony, gritty, sandy loam parent material of this map unit. The parent material, rockiness, slope complexity and climate are very severe limitations on land use which is restricted to rough grazing; such grazings are mainly of low value. Some areas of snow-bed communities and peat flushes do afford limited grazing of moderate value.

The subalpine soils support alpine azalea–lichen heath (Alectorio-Callunetum vulgaris) and lichen-rich boreal heather moor (part of Vaccinio-Ericetum cinereae) with mountain blanket bog (Rhytidiadelpho-Sphagnetum fusci) communities on the peat.

Map unit 13   There are only two areas of map unit 13 in Eastern Scotland (Sheet 5) together totalling 5 square kilometres (2 per cent of the association). The larger of these areas is west of Coignafearn, the other on the plateau north of Glen Markie.

Component soils are peat and subalpine soils with some alpine soils. At Coignafearn in the upper Findhorn valley, the map unit is at 700 metres with an average annual rainfall of 1500 millimetres, whereas at Glen Markie, the unit is at nearly 900 metres and has an average annual rainfall in excess of 2000 millimetres. Both sites are very exposed and have extremely severe winters associated with the very cold, wet mountain climate region.

Locally derived shattered rock forms the major part of the stony, gritty, sandy loam drift on which the soils are developed.

The associated landform is one of mountains with gentle and strong, non-to moderately rocky slopes. The limitations of climate, soils and complex topography are very severe and land use is restricted to rough grazing; grazing values are mainly low, occasionally moderate.

The mountain blanket bog (Rhytidiadelpho-Sphagnetum fusci) is the dominant plant community of the peat areas. On the subalpine and alpine soils, alpine azalea–lichen heath (Alectorio-Callunetum vulgaris) and stiff sedge–fescue grasslands (the *Carex bigelowii–Festuca vivipara* Association) are the common plant communities.

Map unit 14   Although one small area of this map unit occurs to the west of Loch Ericht, the greater part is totally confined to the east–west ridges on either side of Loch an t-Seilich, Gaick. It covers 9 square kilometres, a little over 3 per cent of the association.

The soils are dominantly alpine; they are mainly freely drained and podzolic. Climatically, the areas are very exposed with extremely severe winters and are associated with the very cold wet mountain climate region. The unit is situated above 800 metres and has an average annual rainfall between 1500 and 2000 millimetres.

Mountain summits with gentle or strong, non-rocky and slightly rocky slopes are the landforms. The parent material is shallow and formed from frost-shattered and weathered, local rock.

Very severe limitations of topography, climate and soil restrict land use to rough grazing; common areas of dominant snow-bed vegetation provide grazing of moderate value.

Plant communities are alpine azalea–lichen heath (Alectorio-Callunetum vulgaris), stiff sedge–fescue grasslands (the *Carex bigelowii–Festuca vivipara* Association) and alpine clubmoss snow-bed (the *Lycopodium alpinum–Nardus stricta* Community).

Map unit 15   This map unit covers an area of 19 square kilometres or 7 per cent of the association. Apart from a small area on the Carn Dearg summit in Rannoch Forest, the main locations are the very steep, scree and rock walls of the Tromie valley in Gaick Forest.

Component soils are rankers and lithosols with some alpine soils. In the Tromie valley, the parent material is derived from a complex rock structure of granulite and schist with multiple felsite intrusions. The soils have a very stony, gritty, sandy loam or loamy sand texture and are formed in a colluvium produced from such frost-shattered rock. Bare rock forms a large part of the area.

Because of the physical nature of the unit, the altitude range is wide, 550 to 850 metres, with the average annual rainfall often in excess of 1800 millimetres; the climate regions include the cold wet uplands and the very cold wet mountain.

Mountain summits with strong to very steep and very rocky slopes comprise the landform. Extremely severe limitations, especially soil and climate, restrict the areas on Carn Dearg and on the more exposed, rock-dominated upper slopes around the River Tromie headwaters to seasonal and very poor rough grazings. In the more sheltered slopes of the side valleys the limitations are less severe and the grazings have moderate values.

Plant communities are blaeberry heath (the *Rhytidiadelphus loreus–Vaccinium myrtillus* Community), bog whortleberry heath (the *Racomitrium lanuginosum–Vaccinium uliginosum* Association), alpine azalea–lichen heath (Alectorio-Callunetum vulgaris), and stiff sedge–fescue grasslands (the *Carex bigelowii–Festuca vivipara Association*).

THE ARDVANIE ASSOCIATION

(Map unit 17)

The soils of the Ardvanie Association are developed on morainic drifts derived from Middle Old Red Sandstone strata and Moinian schists. There is only one map unit and its description is included with that of the association. It covers 29 square kilometres or 0.1 per cent of the land area and occurs around Invergordon in Easter Ross and between Inverness and Forres.

The parent material is stony, sometimes stone-dominated, and varies from a gravelly loamy sand to a loamy sandy gravel. Minor bands of sand or, more rarely, of silt occasionally show traces of bedding. Though closely related to the morainic drifts of the Millbuie Association, it is distinguished by the schist content. Locally it may have close affinities with the fluvioglacial gravels of the Corby Association. The colour of the drift is brown or reddish brown.

As with most of the coarse-textured morainic and fluvioglacial sands and gravels, the cation-exchange capacity is low. Calcium, magnesium and potassium values show a sharp decrease in the B horizon and this decrease continues in the C horizon where values range from 0.01 to 0.10 milli-equivalents per 100 grammes, with the majority of values less than 0.05.

The landform consists of undulating lowlands and foothills with occasional areas of mounds. Slopes are normally gentle but are locally steep, especially in the moundy areas. Rock outcrops are rare. In Easter Ross, the altitude ranges from 20 metres near Balintore to around 97 metres in The Wilderness north of Invergordon. South of the Moray Firth, where the unit flanks Drummossie Muir, the altitudes range up to 224 metres. The average annual rainfall is between 600 and 800 millimetres in the lowlands, rising to 1000 millimetres in Strathnairn. Climatically, the unit lies mainly in the warm and fairly warm, moderately dry regions.

Freely drained humus-iron podzols are the dominant soils on the straight or slightly convex slopes on the southern side of the Moray Firth, whereas in the more moundy areas in Easter Ross, freely and imperfectly drained humus-iron podzols are co-dominant. Minor areas of noncalcareous and peaty gleys are associated with sporadic depressional sites, especially within the moundy variant.

Land use is divided, almost evenly, between agriculture and forestry. On the cultivated land, a moderate range of arable crops or a narrow range based on ley pastures (Lolio-Cynosuretum) is found. Where the gleys are not afforested with Sitka or Norway spruce, they are characterized by the soft rush pasture (the *Ranunculus repens–Juncus effusus* Community) and sedge mires. The uncultivated humus-iron podzols support dry Atlantic heather moor (part of Carici binervis-Ericetum cinereae).

THE ARKAIG ASSOCIATION
(Map units 19–23, 25–36)

This association is the second largest within Eastern Scotland (Sheet 5) and covers 3275 square kilometres (13.3 per cent of the land area). Owing to a range of altitude from almost sea level to over 1100 metres and a considerable variation in climate, parent materials, soils and topography, seventeen map units have been used.

The soils are developed on parent materials derived from schists, gneisses, granulites and quartzites, principally of the Moine Assemblage. A stony sandy loam or loam till is widespread in lowland and foothill sites, especially on gentle, straight or slightly concave, slopes. On the upper slopes and hilltops, where shallow drift forms the soil parent material, the texture is normally a stony sandy loam, occasionally loamy sand. At the highest altitudes on mountain summits, the parent material is often a locally derived, shallow, frost-shattered material. Rarely, the parent material consists of deeply weathered rock. Deposits of colluvial material occur on steep slopes, often with a texture of loam or sandy loam and a low stone content. Moraines with a sandy loam or loamy sand texture and a high content of subrounded or rounded stones sometimes form a local parent material on valley floors and valley sides in the uplands. The characteristic bedding, or stratification, of the fluvioglacial material is absent in such deposits.

The association occupies a large area in a wide belt from Rannoch Moor to Inverness.

Most Scottish major soil subgroups are represented in the association. Brown forest soils with free and imperfect drainage are restricted to the lowland sites. Noncalcareous gleys, peaty gleys and humus-iron podzols are common in the lowlands and foothills with peaty podzols at higher elevations. Subalpine soils are widespread above 600 metres with alpine soils occurring normally above 700 metres. On rugged landforms, rankers are extensive.

Nearly all the arable land is capable of producing a moderate range of crops, major limitations being climate, wetness and topography. These factors become more severe in the foothills where arable agriculture is replaced by long ley pasture. The vegetation of the upper hill slopes provides mainly low-quality rough grazing. Land between the present arable limits and the upper hill grazings can often be reseeded successfully to grassland. Slopes below 550 metres have a considerable potential for forestry.

Map unit 19 Although not an extensive map unit, covering only 90 square kilometres, or less than 3 per cent of the association, it has a widespread distribution by virtue of its soil and topographical relationship. Confined to hills and valley sides with gentle and strong, non-rocky slopes, the map unit has a narrow altitude range, 200 to 450 metres, with an average annual rainfall between 750 and 1200 millimetres. Climatically, the map unit is associated with the fairly warm, moderately dry lowlands and with cool, wet foothills and uplands.

The soils are predominantly noncalcareous and peaty gleys with some humic gleys and peat. The parent material is often thicker in the depressions and long gentle concave slopes where it is a stony sandy loam or loam till. Areas occur on strong slopes, however, where there is bed-rock close to the surface and the shallow drift has a coarse sandy loam texture.

In some lowland sites, farming has gradually improved the natural infertility of the soils so that moderate yields of a narrow range of crops are possible. Climate, however, restricts many areas to long ley grassland or permanent grassland with only very limited potential for other crops.

On the cultivated land, permanent and ley pastures (Lolio-Cynosuretum) and arable crops are found, but soft rush pasture (the *Ranunculus repens–Juncus effusus* Community) and sedge mires occur on some of the uncultivated gleys. The peaty gleys and peats have respectively bog heather moors (Narthecio-Ericetum tetralicis) or blanket bogs (Erico-Sphagnetum papillosi) as the dominant plant communities.

Map unit 20 Freely and imperfectly drained humus-iron podzols are the principal soils of this map unit, with some brown forest soils, noncalcareous gleys and peaty gleys. They are often cultivated.

Map unit 20 occupies some 254 square kilometres (or 8 per cent of the association) mainly in Speyside and Deeside. The landform is one of undulating lowlands and valley sides with gentle and strong, non-rocky slopes, generally below 300 metres and rarely above 400 metres. The average annual rainfall has a range of 750 to 1200 millimetres and the unit is associated with the climate regions of fairly warm, moderately dry lowlands and with cool, wet foothills and uplands.

Though the drift may have a stony, sandy loam texture where the rock is close to the surface on strong slopes, the parent material is normally thick with a stony loamy texture. Local areas of sandy clay loam till also occur.

Land use is devoted mainly to moderate and good arable agriculture or to mixed farming with long ley pastures. In areas where more severe climatic or soil restrictions prevail, cultivation may be so hazardous that land use is limited to permanent grassland.

The dominant plant communities are arable crops and permanent and ley pastures (Lolio-Cynosuretum). In uncultivated areas, dry boreal heather moor (part of Vaccinio-Ericetum cinereae) and acid bent–fescue grassland (part of Achilleo-Festucetum tenuifoliae) communities are associated with the humus-iron podzols.

Map unit 21 Within the association this is a minor unit, covering only 92 square kilometres or 3 per cent of the association. Apart from a number of scattered, small areas, it is confined to the foothills of the Dee, Garry and Errochty valleys and around Loch Rannoch. Soils with organic surface horizons dominate the map unit, including peaty podzols, some humus-iron podzols, peaty gleys and peat.

With an altitude range of 200 to 400 metres and an average annual rainfall of the order 750 to 1200 millimetres, the map unit is associated with the climate regions of fairly warm, moderately dry lowlands and with cool, wet foothills.

The landform is one of hills and valley sides with gentle and strong, non-rocky slopes. The parent material is a stony, sandy loam drift which is often shallow on strong slopes.

In some areas where the organic surface horizon is not too thick, the soils have been cultivated though land use is restricted to long ley grassland or permanent pastures. Generally, the organic surface is so thick that the land is marginally suited only to grassland reclamation and land use is limited to rough grazing; the grazings are mainly of low value.

On the podzolic soils, plant communities are dry and moist boreal heather moors (parts of Vaccinio-Ericetum cinereae), whereas on the peat and peaty gleys, bog heather moors (Narthecio-Ericetum tetralicis) and blanket bogs (Erico-

Sphagnetum papillosi) dominate the vegetation. In cultivated areas, permanent and long ley pastures (Lolio-Cynosuretum), soft rush pasture (the *Ranunculus repens–Juncus effusus* Community) and sedge mires occur.

Map unit 22   This is the most extensive map unit of the Arkaig Association, occupying some 413 square kilometres or nearly 13 per cent of the association. It is widespread throughout the association area, but occurs especially in the foothills and uplands around Strathdearn and Speyside. Component soils are peaty podzols and peat with some peaty gleys and humus-iron podzols. Hills and valley sides with strong, non-rocky slopes are the landforms.

The parent material is a stony, sandy loam drift which is often shallow on the strong slopes and hill tops. Where the slopes are not too strong and are long and bench-like, however, a till of a sandy loam, locally a loam, texture forms the parent material.

The altitude and average annual rainfall ranges are wide, from 200 to 600 metres and 800 to above 1600 millimetres respectively. Climatically, the map unit is associated with the regions of the fairly warm, moderately dry lowlands, the cool, wet foothills and uplands, and the cold, wet uplands.

Limitations to land use are poor or very poor surface drainage, thick organic horizons, climate and topography. Usually the land is moderately suited to grassland reclamation, but where surface water is excessive and where the organic horizon is thick, it is only marginally suited. Where improvement is not practicable or is difficult to sustain, for example, on the peat or peaty gleys, land use is restricted to rough grazing.

Moist boreal heather moor (part of Vaccinio-Ericetum cinereae) is the main plant community on the peaty podzols and humus-iron podzols. On peat and where the organic horizon thickens on peaty gleys, lowland and upland blanket bogs (parts of Erico-Sphagnetum papillosi) and bog heather moors (Narthecio-Ericetum tetralicis) are dominant.

Map unit 23   This unit covers over 10 per cent of the association, 332 square kilometres and is similar to *map unit 22* but wetter; it shares a widespread distribution. The landform is one of undulating lowlands and uplands with gentle and strong, non-rocky slopes.

Wet soils are dominant, mainly peaty gleys and peat with some peaty podzols. The parent material is generally a thin, stony, sandy loam drift, although in some of the depressions or bench-like areas a thick, stony, sandy loam till, locally a loam, is present.

Altitude and average annual rainfall ranges are wide, 200 to 600 metres and 850 to 2000 millimetres respectively. The climate regions are cool, wet foothills and uplands and cold, wet uplands.

Land use is restricted mainly to rough grazing because of very severe limitations including wetness, climate and organic surface horizons; the grazings are of low or moderate value. A few limited areas are marginally suited to grassland reclamation. Windblow may be a hazard to forestry on the higher slopes.

Common plant communities on the peat and peaty gleys are lowland and upland blanket bogs (parts of Erico-Sphagnetum papillosi) and bog heather moors (Narthecio-Ericetum tetralicis). On the peaty podzols there is often the moist boreal heather moor community (part of Vaccinio-Ericetum cinereae).

Map unit 25   It is one of the smaller units, occupying only 62 square kilometres, a little under 2 per cent of the association. Of limited distribution, it occurs mainly along the Braes of Atholl, Strath Tummel and Strath Errochty. The soils are mostly brown forest soils and humus-iron podzols with classical examples of the latter found in areas of the Caledonian pinewood. In small areas on the shores of Loch Rannoch, the brown forest soils are transitional to podzols. Minor soils are gleys and peat.

The landform is a hummocky valley moraine which often has a bouldery surface. The parent material is a stony, often very stony, gritty sandy loam or loamy sand. Although most of the moraines have had only a limited degree of water-sorting many of the stones are considerably rounded. Some moraines, however, are formed from locally derived subangular material.

The altitude range is narrow, 200 to 450 metres, but is generally less than 400 metres. The average annual rainfall is in the range 900 to 1200 millimetres. Climatically, the areas are within the fairly warm, moderately dry lowlands and the cool, wet foothills and uplands climate regions.

In most situations, the coarse texture and stoniness, combined with the restriction of the moundy topography, limit the land use to permanent grassland or long ley grassland. Where slopes are steep and the ground pattern is very complex, land use is limited to rough grazing; the grazings have moderate or low relative values.

Plant communities include permanent pastures (Lolio-Cynosuretum), acid bent–fescue grassland (part of Achilleo-Festucetum tenuifoliae) and Atlantic oakwood and western highland birchwood (Blechno-Quercetum) and native pinewood (Pinetum scoticae).

Map unit 26   This map unit covers 236 square kilometres or nearly 7 per cent of the association. It is closely related to *map unit 25* but differs in that the surface horizons are peaty. Dominant soils are peaty podzols, peat and peaty gleys, and the map unit is widespread in upland valleys.

The parent material is a stony, often very stony, sandy loam, or loamy sand, moraine. As with *map unit 25*, the landform is one of a hummocky valley and slope moraine, which is often bouldery (Plate 8). The altitude range of 350 to over 600 metres is greater than that for *map unit 25* and the average annual rainfall is rarely below 1000 millimetres and generally exceeds 1400 millimetres. Climatically, the unit is associated with the cool, or cold, wet foothills and uplands. Areas above 500 metres are generally within the cold, wet uplands climate region. The very severe limitations of poor drainage, thick organic surface horizons, complex topography and climate, restrict the land use to rough grazing; the grazings are of low, occasionally moderate, values.

Where the organic horizon is not too thick, for example, on the peaty podzols, the plant communities are moist boreal and Atlantic heather moors (part of Vaccinio-Ericetum cinereae and Carici binervis-Ericetum cinereae respectively). On the peat and peaty gleys, lowland and upland blanket bogs (parts of Erico-Sphagnetum papillosi) and bog heather moors (Narthecio-Ericetum tetralicis) dominate.

Map unit 27   This map unit is not extensive, covering only 52 square kilometres or 2 per cent of the association. It is confined to the valleys of Lochs Ericht, Ness, Rannoch and Tummel and Glens Tarff, Killin and Tilt where the valley sides have steep or very steep, moderately rocky and very rocky slopes.

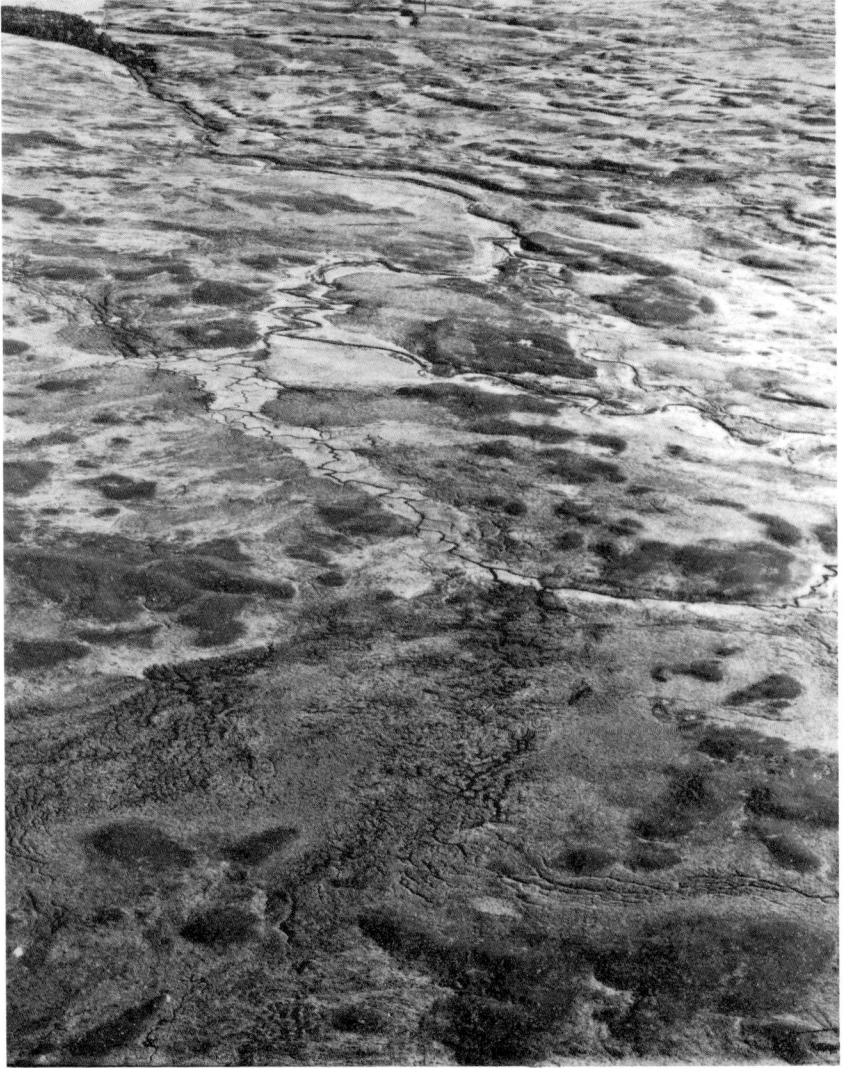

Plate 8. *South of Loch Rannoch near Dall Burn, showing morainic mounds and peat hollows. The peaty podzols, peaty gleys and peat of map unit 26 (Arkaig Association, Class 6) are extensive. Small areas of mineral and peaty alluvial soils are confined to the stream sections.* SDD Crown copyright.

Component soils are brown forest soils and humus-iron podzols with some gleys and rankers. The parent material is a stony sandy loam drift of variable thickness, often including much locally derived, subangular and angular fragmented rock. Associated with the base of the steeper slopes, there is a finer-textured stony colluvium where brown forest soils or brown podzolic soils are developed. The constant colluviation of the upper horizons inhibits the development of peaty surface horizons.

The altitude range is wide, 50 to 450 metres and the average annual rainfall is of the order 1000 to 1400 millimetres. Climatically, the map unit is in the region of cool, wet foothills and uplands; often the limitations of slope and rockiness are very severe, for example, Glen Tilt, and land use is restricted to rough grazing; much of the grazings has a moderate relative grazing value. Where, the slope and rock limitations are less severe, the land is moderately suited to grassland reclamation, for example, around Loch Tummel.

Plant communities include rich bent–fescue grassland (part of Achilleo-Festucetum tenuifoliae) on the brown forest soils, and dry boreal and Atlantic heather moors (parts of Vaccinio-Ericetum cinereae and Carici binervis-Ericetum cinereae, respectively) on the humus-iron podzols. There are occasional mixed broadleaved woodlands mainly Atlantic oakwood and western highland birchwood (Blechno-Quercetum).

Map unit 28   Covering 368 square kilometres or approximately 11 per cent of the association this is a unit of the northern and western sectors with an altitude range of 250 to over 700 metres and an average annual rainfall of 750 to 1600 millimetres. Climatically, the map unit is within the cool, wet foothills and uplands or cold, wet uplands.

East of the Great Glen, the dominant soils are peaty podzols and humus-iron podzols with some peaty gleys and rankers. The parent material is a shallow, stony sandy loam drift which includes variable quantities of locally derived, subangular and angular rock fragments. The landform is one of hills and undulating lowlands with gentle and strong, moderately rocky slopes. West of the Great Glen, the unit is often wetter with more peaty gleys and also some peat, the typical eastern unit being confined usually to steep slopes.

Often the rockiness, organic surface horizons and slope complexity are very severe limitations and restrict land use to rough grazings, which have moderate and low relative grazing values. Much of the unit is, however, well suited to timber production and many areas have been afforested. Successful grassland reclamation is possible on areas where humus-iron podzols dominate and where the rock outcrops are sufficiently widely spaced.

Plant communities are boreal and Atlantic heather moors (Vaccinio-Ericetum cinereae and Carici binervis-Ericetum cinereae respectively), heath rush–fescue grasslands (Junco squarrosi-Festucetum tenuifoliae) and rich bent–fescue grassland (part of Achilleo-Festucetum tenuifoliae).

Map unit 29   This unit is similar to *map unit 28* but is wetter and covers an area of 196 square kilometres or 6 per cent of the association. The soils are peaty gleys and peat with some peaty podzols and peaty rankers. The parent material is a stony, often shallow, sandy loam drift.

The altitude and average annual rainfall ranges are wide, 300 to over 800 metres and 1000 to over 2000 millimetres respectively. Several climatic regions are associated with this unit, including mainly cool, wet foothills and uplands and cold, wet uplands and with the highest areas in the very cold, wet mountains. The landform is one of undulating hills with gentle and strong, moderately rocky slopes.

As with *map unit 28*, the very severe limitations of wetness, organic surface horizons, complex slopes and rockiness restrict land use to rough grazings; grazing values are low, occasionally moderate. Areas could be afforested but windthrow may be a hazard.

Plant communities are mainly bog heather moors (Narthecio-Ericetum tetralicis) and lowland and upland blanket bogs (parts of Erico-Sphagnetum papillosi) with moist boreal heather moor (part of Vaccinio-Ericetum cinereae) on the drier soils.

Map unit 30   This map unit is widespread but is especially extensive on the drier hills west of the River Spey. It is similar to *map unit 28* but is rock dominated and covers an area of 218 square kilometres, approximately 7 per cent of the association.

Component soils are peaty podzols and rankers with some humus-iron podzols and peaty gleys. Apart from very local deposits of lodgement till, the soil parent material is a stony, often very stony, sandy loam drift. A varying proportion of locally derived, fragmented rock is often present. On steeper slopes, some colluviation has taken place, forming a slightly finer-textured parent material towards the base of the slope. The landform is one of rugged hills with strong and steep, very rocky slopes.

The average annual rainfall is mainly within the range 750 to 1800 millimetres. The altitude range is very wide, from 50 to over 700 metres. Climatically, areas of map unit 30 occur mainly in the fairly warm moderately dry lowlands, the cool wet foothills and uplands and the cold wet uplands.

Land use is very severely restricted by the complex slope pattern and rockiness, usually affording only rough grazing which has a low value. The very rocky areas, however, can sustain only a very limited, poor rough grazing. The unit has been afforested in some areas and may be reasonably productive, but extraction of the timber is hazardous.

Plant communities are dry and moist boreal heather moors (parts of Vaccinio-Ericetum cinereae) on the peaty podzols, humus-iron podzols and rankers. Bog heather moors (Narthecio-Ericetum tetralicis) dominate both the peaty gleys and peaty podzols with thicker organic surface horizons. Blaeberry heath (the *Rhytidiadelphus loreus–Vaccinium myrtillus* Community) is occasionally present.

Map unit 31   This is a minor map unit, occupying only 52 square kilometres or approximately 2 per cent of the association. It is commonly associated with the wet, western Highlands and is found only along the boundary with Western Scotland (Sheet 4), north of Corrour. There, it is locally widespread in Glendoe Forest and around the headwaters of the River Spey. The dominant soils are peaty gleys, peaty podzols and peaty rankers.

Hill sides with steep and very steep, moderately and very rocky, slopes are the main landforms. The soil parent material is a very stony, sandy loam shallow drift with a varying proportion of locally derived, fragmented rock; a finer-textured colluvium is often present at the base of steep slopes.

With an altitude range of 300 to over 800 metres and an annual average rainfall of 1400 to 2000 millimetres, climatically the unit occurs within the cold, wet uplands and the very cold, wet mountain regions. Major limitations of topography, climate and rockiness restrict the land use to rough grazings; the common flushed sites provide grazings of moderate value otherwise values are low.

Plant communities are bog heather moors (Narthecio-Ericetum tetralicis), Atlantic heather moors (Carici binervis-Ericetum cinereae) and acid bent–fescue grassland (part of Achilleo-Festucetum tenuifoliae).

Map unit 32   Occupying only 13 square kilometres or less than 1 per cent of the association, this map unit is the smallest Arkaig unit in Eastern Scotland (Sheet 5). The main areas are confined to the north of Invermoriston on the west side of Loch Ness. It is a map unit more commonly found in the wet western hills and is similar to *map unit 30* but is wetter. The landform is one of rugged hills with gentle and strong, very rocky slopes.

The parent material is a stony, shallow, sandy loam drift, often with locally derived, subangular or angular rock fragments. Component soils are peaty gleys, peaty rankers and peat. The altitude range is from 400 to over 600 metres and the average annual rainfall is 1400 to 2000 millimetres.

Climatically, the map unit is mainly in the cold, wet uplands. Land use is restricted generally to rough grazing because of topography, rockiness and wetness. Grazing values are low with the very rocky areas providing only limited grazing.

Plant communities include moist boreal and Atlantic heather moors (part of Vaccinio-Ericetum cinereae and Carici binervis-Ericetum cinereae respectively), bog heather moors (Narthecio-Ericetum tetralicis), and lowland and upland blanket bogs (parts of Erico-Sphagnetum papillosi).

Map unit 33 This is the most extensive of the upland and mountain units, covering 377 square kilometres, approximately 12 per cent of the association. It is widespread throughout the association area. Component soils are subalpine soils with some peat, rankers and alpine soils. The soil parent material is a shallow, stony, sandy loam drift. This drift in some locations is absent or very shallow and gives way to a locally derived parent material of frost-shattered, subangular or angular rock.

The landform is one of mountains with gentle to very steep, non- to very rocky slopes. With an average annual rainfall in excess of 1600 millimetres and an altitude from 550 to above 800 metres, the map unit belongs to the cold wet uplands and the very cold, wet mountain climate regions.

Restrictions on land use are topography, rockiness, soil, wetness and climate. These limitations are very severe and afford only rough grazings; they have mainly low, occasionally moderate, grazing values. The rock-dominated areas provide very limited, seasonal, poor grazing whereas areas of snow-bed vegetation and flushed soils provide grazing of moderate value.

The main plant communities are alpine azalea–lichen heath (Alectorio-Callunetum vulgaris), lichen-rich boreal heather moor (part of Vaccinio-Ericetum cinereae) and mountain blanket bog (Rhytidiadelpho-Sphagnetu fusci).

Map unit 34 Map unit 34 is associated with the high-level zones. Though its distribution is dependent on altitude, it is widespread, covering 200 square kilometres, approximately 6 per cent of the association. Individual large areas are especially common in the Monadhliath Mountains. Component soils are peat and subalpine soils with some alpine soils. The parent material is a stony, shallow, sandy loam drift. The drift is of very variable thickness and often gives way to a locally derived, frost-shattered rock.

Altitude is generally above 800 metres and rainfall is in excess of 1600 millimetres. Climatically, the unit is associated mainly with the very cold wet mountain region. The landform is one of mountains with gentle and strong, non- to moderately rocky slopes. Restrictions of climate, exposure, soil, wetness and topography are very severe and limit land use to rough grazings; these have a low, occasionally moderate, grazing value.

Plant communities include mountain blanket bog (Rhytidiadelpho-Sphagnetum fusci) on the peat, and alpine azalea–lichen heath (Alectorio-Callunetum vulgaris) and stiff sedge–fescue grasslands (the *Carex bigelowii–Festuca vivipara* Association) on the mineral soils.

Map unit 35   This high-level mountain unit covers an area of 140 square kilometres or a little over 4 per cent of the association. It is fairly widespread and occurs on mountain summits with gentle and strong, non- and slightly rocky slopes above 800 metres. Rainfall exceeds 1800 millimetres and climatically the unit belongs to the very cold, wet mountain region. The parent material is usually frost-shattered, mountain-top detritus though occasionally the soil may be developed on completely weathered rock. Component soils are alpine soils with some rankers. Limitations of climate, exposure, soil, wetness and topography are very severe and restrict land use to rough grazings. Where grassy snow-bed vegetation is dominant, grazing of a moderate value is available, for example, on the hills east of Dalwhinnie, otherwise values are low.

Plant communities include alpine azalea–lichen heath (Alectorio-Callunetum vulgaris), stiff sedge–fescue grasslands (the *Carex bigelowii–Festuca vivipara* Association) and alpine clubmoss snow-beds (the *Lycopodium alpinum–Nardus stricta* Community).

Map unit 36   This map unit covers an area of 178 square kilometres, approximately 5 per cent of the association, and is widespread throughout the association. There are two characteristic landforms; both are very rocky. A very steep form is confined to mountains with well-developed corries or complex scree slopes, for example, Ben Alder, and to the walls of deeply incised valleys for example, Glen Markie; a horizontal form relates to mountain summits with extensive areas of bare pavement rock and boulder lobes. Both forms occur often within the same area, but the proportion of each varies considerably.

On the summits, the soil parent material is locally derived, frost-shattered rock debris whereas on the valley sides the soils are developed in scree and colluvium. Component soils are rankers and lithosols; some alpine soils occur occasionally on the mountain tops.

Because of the physical nature of the unit, the altitude range is wide, from 450 metres to over 900 metres, and the average annual rainfall is generally above 1500 millimetres. The areas are associated with the cold wet uplands and the very cold wet mountain climate regions. Land use restrictions are very severe and extremely severe; only limited seasonal grazings of a low value are usually available on the mountain and corrie sites whereas grazings of moderate and low values characterize the valley sides.

Plant communities on the mountain tops are blaeberry and bog whortleberry heath (the *Rhytidiadelphus loreus–Vaccinium myrtillus* Community and the *Racomitrium lanuginosum–Vaccinium uliginosum* Association), and alpine azalea–lichen heath (Alectorio-Callunetum vulgaris). Stiff sedge–fescue grasslands (the *Carex bigelowii–Festuca vivipara* Association) is locally common on snow-beds. The valley walls are dominated by upland bent–fescue grassland and blaeberry heath (part of Achilleo-Festucetum tenuifoliae and the *Rhytidiadelphus loreus–Vaccinium myrtillus* Community respectively).

THE BALROWNIE ASSOCIATION

(Map units 41–46)

The soils of this association are developed on drifts, often water-modified, which are derived from Lower Old Red Sandstone sediments containing both coarse-grained massive sandstones and fine-grained flaggy sandstones. Covering a total area of 1239 square kilometres (5.0 per cent of the land area), the association occurs extensively throughout Strathmore and Strathallan where it is confined largely to ground at altitudes between 60 and 120 metres. On the Braes of Doune in Perthshire and on parts of the Sidlaw Hills in Angus, however, it extends to over 300 metres on some upper slopes but is mainly below 210 metres. The drift, which is generally reddish brown in colour, varies in texture from fine sandy loam to clay loam and has a moderate stone content of sandstones with occasional schist erratics. Over much of the low ground in Strathmore and Strathallan, the upper layers of the drift have been considerably modified by periglacial activity, including some degree of water-sorting associated with the melting of the last main ice-sheet. As a result, the soils frequently have partially sorted sandy loam or loam upper horizons overlying compact sandy clay loam or clay loam drift. The Balrownie Association occurs in climate regions ranging from fairly warm, moderately dry lowlands with an average annual rainfall of 700 millimetres to fairly warm, wet lowlands and foothills with 2000 millimetres rainfall.

Map unit 41   This is the most extensive map unit in the Balrownie Association and occupies 948 square kilometres (76 per cent of the association) in Strathmore, in east Perthshire and in parts of the Sidlaw Foothills. It consists of imperfectly drained brown forest soils with gleying, together with some freely drained brown forest soils and some poorly drained noncalcareous gleys. The dominant soil has a loam topsoil overlying a sandy loam or loam subsoil, frequently in partially sorted material, which passes into sandy clay loam or clay loam drift.

Occurring on non-rocky undulating lowlands and foothills with gentle and strong slopes over an altitude range of 60 to 120 metres, the soils of this unit are mainly cultivated; some areas are under coniferous or mixed woodland, the latter being mostly elmwood (Querco-Ulmetum glabrae) and southern oakwood (Galio saxatilis-Quercetum). The arable land produces mainly a wide range of crops with high yields. Some areas, however, produce only a moderate range with yields varying from high to average.

Map unit 42   This unit is not extensive, covering only 97 square kilometres (8 per cent of the association). It occurs in depressions and on gentle slopes in the lowlands of Strathmore and east Perthshire, and comprises noncalcareous gleys with some peaty and humic gleys. Much of the unit is cultivated; some areas are in permanent pasture (Lolio-Cynosuretum) whereas the more humic soils support rush pasture or sedge mires. Minor areas of broadleaved woodland consist of elmwood (Querco-Ulmetum glabrae) and southern oakwood (Galio saxatilis-Quercetum). The cultivated land produces a moderate range of arable crops with average yields though some areas are restricted to a narrow range based primarily on grassland. The uncultivated ground is suited for grassland reclamation.

Map unit 43   Consisting of freely drained brown forest soils with small areas of imperfectly drained brown forest soils, it is similar to *map unit 41* but differs in the nature of the parent material and in the resulting soil textures. It covers 82 square kilometres (7 per cent of the association). Developed on shallow till or on residual sandstone rock, the soils are found in areas of undulating lowland with gentle and strong slopes. There are occasional rock outcrops. The soil profile has a fine sandy loam or sandy loam topsoil, sometimes developed in water-sorted material, overlying a sandy loam or loamy sand layer, often indurated, on loamy coarse sand containing pieces of decomposing sandstone. Capping and poaching may be problems in the fine sandy loam topsoils. Much of the unit is in pasture or rough grazing but is suitable for grassland reclamation. Some areas are under mixed woodland, mainly elmwood (Querco-Ulmetum glabrae) and southern oakwood (Galio saxatilis-Quercetum). Within the cultivated land, most of the arable areas are restricted to a moderate range of crops with average yields. A few minor areas do produce, however, a wide range with high yields.

Map unit 44 This unit consists of humus-iron podzols with some brown forest soils and peaty podzols and is found in foothills and valley sides in the Ochil and Sidlaw Hills. It is a minor unit and covers only 57 square kilometres (5 per cent of the association). The soils, which are developed on drift of sandy loam or loam texture, are generally non-rocky and occur on gentle and strong slopes. Much of the area is afforested but is otherwise suited to grassland reclamation.

Map unit 45 Comprising peaty podzols, some humus-iron podzols, peat and peaty gleys, it is found mainly in the Cromlix and Tullibardine areas to the west of Auchterarder. Developed on loamy drift, the soils are non-rocky and occur on undulating uplands with gentle and strong slopes. The unit is in pasture or rough grazing and covers 26 square kilometres (2 per cent of the association). There is potential for grassland reclamation.

Map unit 46 Developed on sandy clay loam drift this unit consists, dominantly, of peaty gleys with some peat and occasional peaty podzols. It occurs on undulating lowlands and uplands with gentle slopes and is of limited extent, covering 29 square kilometres (2 per cent of the association). Land use is mainly rough grazing and forestry though there is scope for grassland reclamation. Climate and wetness are the restricting factors.

THE BERRIEDALE ASSOCIATION

(Map unit 61)

The soils of the Berriedale Association are derived from sandstones and conglomerates of the Middle Old Red Sandstone Barren Group. Some of this Middle Old Red Sandstone has now been reclassified as Lower Old Red Sandstone (Institute of Geological Sciences, 1979). In this area the parent material is a stony till derived mainly from sandstone.

It is one of the least extensive associations in Eastern Scotland (Sheet 5) and is found only in the extreme north-west corner of the map. It covers less than 1 square kilometre but is only the eastern extremity of a larger area which occurs in Northern Scotland and Western Scotland (Sheets 3 and 4). The association occurs almost exclusively in Northern Scotland where this map unit is by far the most extensive of the nine units. As there is only one map unit of the association in Eastern Scotland, however, its description has been combined with that of the association.

The landform consists mainly of gently sloping hill sides. The average annual rainfall is around 1600 millimetres and the area occurs in the cold, wet foothill and upland climate region.

The soils are dominated by peaty gleys and peat because of the effect of the wet climate on the gentle slopes. Peaty podzols are subsidiary in this area. Bog heather moors (Narthecio-Ericetum tetralicis) occur on the peaty gleys, whereas lowland and northern blanket bog (parts of Erico-Sphagnetum papillosi) are found on the peat.

THE BOGTOWN ASSOCIATION

(Map unit 70)

This is one of the smallest associations in Eastern Scotland (Sheet 5), covering 10 square kilometres. There is only one map unit and its description is included with that of the association. It occurs mainly between Cullen Bay and Boyne Bay on the Banffshire coast on smooth, gentle slopes below 61 metres; an isolated, minor area occurs near Troup Head.

The parent material is a pink to light reddish brown lacustrine deposit which varies from silt to clay; most textures are silty clay. It is commonly over 1 metre thick and usually overlies black Jurassic clay or water-sorted sands and gravels. Buried surface horizons are common and the deposit is thought to be interglacial (Read, 1923).

The soils are but occasionally very poorly drained calcareous and non-calcareous gleys, mostly poorly drained. Upper horizons have a strong prismatic structure, but the lower horizons are massive. A few stones occur in the surface horizon.

The average annual rainfall is around 750 millimetres and the unit lies in the climate regions of warm and fairly warm moderately dry lowlands.

Arable agriculture is the dominant land use. The whole area is capable of producing a moderate range of crops with high or moderate yields. This is the direct consequence of centuries of drainage and improvement.

THE BRAEMORE/KINSTEARY ASSOCIATIONS

(Map units 71–74)

The Braemore and Kinsteary Associations, both comprising soils developed on drifts derived from argillaceous rocks of mainly Lower Old Red Sandstone age, have been combined on the 1:250 000 soil map.

Two types of parent material are present. The first is a greyish brown till derived from mudstones and shales; textures are mainly clay loam or silty clay loam. This till is generally found on the gently sloping lower ground. The second type is shallower, usually with a sandy loam or silty loam texture, although silty clay loam and clay loam occur where mudstone is the parent rock. This drift is usually present on the strongly sloping higher ground. Compared with soils derived from the surrounding sandstone, conglomerate and schist, the soils of these associations have much finer textures. Most of the stones in the profile are small.

The combined association is one of the less extensive in Eastern Scotland, covering 30 square kilometres or 0.1 per cent of the land area, and occurs around Dingwall and to the south-west of Auldearn.

The landscape is dominantly non-rocky and gently undulating, with a general altitude range of 150–300 metres. The low ground has an average annual rainfall of 800 millimetres and occurs in the climate region of fairly warm, moderately dry lowland; the higher ground has a rainfall of 1200 millimetres and occurs in the cool wet lowlands and foothills.

Humus-iron podzols are most common, although brown forest soils and non-calcareous gleys also occur. The podzols are usually imperfectly drained with mottled B horizons due largely to fine texture and the frequent proximity of bed-rock to the surface. The brown forest soils are usually shallow and both freely and imperfectly drained soils occur. Reddish brown colours predominate, with the redness becoming stronger with depth. To the south-west of Strathpeffer these soils are developed locally on till derived from grey shales with calcareous bands. The noncalcareous gleys are developed on till derived from shale and occur on the gently sloping lower ground; the fine textures and gentle slopes assist the development of strong gleying. Humic gleys often occur in close association with the noncalcareous gleys.

On the cultivated land, permanent and ley pastures (Lolio-Cynosuretum) and arable crops are found, but soft rush pasture (the *Ranunculus repens–Juncus effusus* Community) and sedge mires occur on some of the uncultivated gleys. The humus-iron podzols support dry Atlantic heather moor (part of Carici binervis-Ericetum cinereae) and, less commonly, acid bent–fescue grassland (part of Achilleo-Festucetum tenuifoliae). Boreal heather moors (Vaccinio-Ericetum cinereae) are also present in the north-west, possibly reflecting the less oceanic climate in the lee of Ben Wyvis. A large proportion of this land is afforested, mainly Sitka and Norway spruce (*Picea* plantations).

There are four soil map units in the Braemore/Kinsteary Associations.

Map unit 71 This unit consists of brown forest soils and it occurs on non-rocky valley sides with gentle and strong slopes. The soils are developed on the shallower drift and on strongly weathered rock. The unit occurs in the Strathpeffer district and covers 7 square kilometres (25 per cent of the combined associations).

Most of the unit consists of freely and imperfectly drained brown forest soils on long regular slopes, but to the south-west of Strathpeffer, the landscape is more rolling. Brown forest soils are still dominant, however, and occur on the hillocks with gleys and, locally, peat in the associated hollows.

Map unit 72 This is a unit of humus-iron podzols, podzolic rankers, humic gleys and some noncalcareous gleys, and it occurs on moderately rocky undulating hill sides with gentle and strong slopes. The soils are developed on till and a shallow stony drift. The unit covers only 1 square kilometre (less than 5 per cent of the combined associations) in the Clach Liath Forest, north of Dingwall.

Freely and imperfectly drained podzols and rankers occur on parallel rock ridges, with the flushed gleys in the associated channels and hollows. The land is suitable only for pasture improvement because of pattern limitations imposed by rockiness and wetness.

Map unit 73 This unit consists of noncalcareous gleys developed on till and it occurs on non-rocky valley sides with gentle slopes. The only occurrence is in Strath Sgitheach where the unit covers 8 square kilometres (25 per cent of the combined associations).

The unit occurs on the lower concave slopes of the valley and although it is non-rocky, shale is found occasionally within 50 centimetres of the surface. Flushing is common on parts of the slope.

In the lower parts of the strath, wetness limitations restrict land use mainly to permanent pastures with limited arable rotation. At higher altitudes, climatic limitations restrict agricultural use and much of the land is under forestry.

Map unit 74 This unit is dominated by humus-iron podzols which are developed mainly on a shallow till. The landform consists of non-rocky hills and undulating lowlands with gentle and strong slopes. The unit covers 14 square kilometres (45 per cent of the combined associations) and is most extensive immediately north of Dingwall. Smaller areas occurs near Ardross and Auldearn.

In the largest area, the unit occurs on a broad ridge which slopes towards the north-east. Apart from the lower southern slopes, most of the ground is under forestry or uncultivated. Slope limitations on the ridge and climatic limitations on the gently sloping summit restrict land use mainly to pasture improvement. By contrast in the Ardross area, where the till is much thicker and accompanied by a slightly more favourable climate, these soils are cultivated with the emphasis on grassland.

THE BRIGHTMONY ASSOCIATION

(Map unit 76)

The Brightmony Association is one of the smaller associations in Eastern Scotland (Sheet 5) and is found only on the southern side of the Moray Firth between Inverness and Forres. It occupies 21 square kilometres or less than 0.1 per cent of the land area. Because there is only one map unit, its description has been included with that of the association.

Interbedded sands and gravels form the parent material, with textures ranging from slightly gravelly loamy fine sands to sandy gravels, the former being dominant. The stones vary widely in their lithology, but are mainly sandstones with common schists and granites, usually rounded or subrounded and less than 5 centimetres in diameter. A pinkish or reddish cast is normally associated with the sand bands. Reddish brown sandy loam till derived from Upper or Middle Old Red Sandstone strata underlies the interbedded drift at depths of 120 to 200 centimetres. Characteristic polygonal frost cracks mark the upper surface of the till.

The non-rocky landscape consists mainly of long broad ridges with gentle slopes within an altitudinal range of 10 to 76 metres. A minor element comprises a deeply dissected slope, to the east of Cawdor, which rises to approximately 140 metres; short strong slopes lead to misfit streams in narrow alluvial channels. The average annual rainfall of 600 to 800 millimetres is one of the lowest in Scotland and the climate is warm and moderately dry.

Podzols, mainly freely drained humus-iron podzols with some iron podzols, are the dominant soils. Induration and an associated well-developed platy structure are characteristic of the B horizon. Minor areas of gleys occur around Ordhill to the south of Auldearn.

Apart from small areas of mixed deciduous and coniferous woodland at Darnaway and Ordhill, the association is devoted to arable agriculture and produces mainly a moderate range of crops. Much of the area is concerned with barley production.

73

THE CARPOW/PANBRIDE ASSOCIATIONS

(Map unit 89)

The two associations are grouped together as both are developed on raised beach sands and gravels derived mainly from Old Red Sandstone igneous and sedimentary rocks. Soils of the Carpow Association are developed on raised beach material and on upper terrace river deposits, whereas those of Panbride are all formed on raised beach deposits. Together, they cover a total area of 129 square kilometres (0.5 per cent of the land area). Although the soils of the Panbride Association have been classed as a podzols, and evidence of podzolization can be seen in the few remaining areas of uncultivated land, they closely resemble brown forest soils as a result of long-term cultivation.

Only one unit, map unit 89, has been distinguished. It consists of freely drained brown forest soils, some imperfectly drained brown forest soils with gleying and cultivated podzols. The profile has, generally, a surface horizon of sandy loam or loam texture overlying a sandy loam or loamy sand subsoil. This latter layer becomes coarser below about 40 centimetres and eventually passes into sandy gravel. Topsoil depths vary from 30 centimetres on most of the inland soils to 50 centimetres, or sometimes more, on some coastal soils. These deep topsoils have resulted, in most cases, from the build-up over the years of organic matter, largely by the application of seaweed.

The unit is found in climate regions of warm moderately dry lowland with an average annual rainfall of 700 to 760 millimetres. Occurring on level to gently sloping terraces at altitudes from 10 to 35 metres, some of the most productive arable soils in Eastern Scotland (Sheet 5) are located between Arbroath and Carnoustie. They produce a very wide range of crops. Other arable areas, found along the lower valleys of the Rivers Tay and Eden, in Stratheden and on the Fife coast, produce a wide range of crops but are unsuited to winter-harvested crops. Scattered throughout the map unit are minor areas of elmwood (Querco-Ulmetum glabrae).

THE CORBY/BOYNDIE/DINNET ASSOCIATIONS

(Map units 96–98, 100 and 101)

The soils of the Corby, Boyndie and Dinnet Associations have been amalgamated for the 1:250 000 map. Those of the Corby Association are developed on fluvioglacial and raised beach sands and gravels derived from acid schists and granite. In some localities adjacent to the Highland Boundary Fault, the parent material contains gravel derived from Old Red Sandstone lavas and dolerites etc. Other rocks from Old Red Sandstone strata are incorporated locally around the Moray Firth. Because of their similar nature, soils of the Dinnet Association have been included. Commonly associated with the gravel deposits are extensive spreads of sand with little gravel admixture. These sands are assigned to the Boyndie Association.

Textures within the Corby Association tend to be coarse, varying from sandy loam or loamy sand in the surface horizons to coarse sandy gravel in the subsoil. Less than 2 per cent clay (less than 2 μm) and less than 5 per cent silt (2–50 μm) are present in the relatively unaltered parent material. In these stratified deposits, excluding thin seams of pure sand, the stone content is very high and consists mainly of rounded and subrounded stones of varying sizes. The thickness of these deposits varies enormously, ranging from about 1 metre to more than 100 metres. In the Boyndie Association, the parent material is characterized by a very high sand content, 90–95 per cent, in the subsoil though in a few localities, the sand is very fine and has approximately 20 per cent (2–50 μm) silt. Resulting from a glacial readvance, the coarse-textured parent material of the Dinnet Association has a high proportion of water-worn cobbles, gravel and coarse sand. The clay content is generally less than 10 per cent.

Both the Corby and Boyndie Associations occur mainly on broad outwash plains and high raised beaches; they are located also on river terraces and in areas of dead-ice topography. The Dinnet Association is restricted to the middle reaches of the River Dee valley. The combined associations occupy 1871 square kilometres or 7.6 per cent of the land area of Eastern Scotland (Sheet 5).

On such acid and coarse-textured parent materials, a wide range of podzols occurs. Although freely drained, natural humus-iron podzols are found on the warm, moderately dry lowlands, most podzols are cultivated. Natural iron podzols and humus podzols are similarly of limited occurrence. Brown forest soils are absent except where the influence of man has retarded podzolization, for example, in planted oakwood or in areas of previous cultivation now identified by the presence of birchwood. Poorly drained soils tend to be confined to wet depressions. With increased altitude, rainfall rises and temperature falls, a combination promoting the accumulation of organic matter and hence the development of peaty podzols. Though some of these peaty podzols are cultivated, the natural soils are widespread on open moorlands and on gentle hill slopes. These podzols closely resemble a variant of the humus-iron podzol found in pinewoods in which the felty organic horizons, typical of the forest environment, are replaced by a thin horizon of raw humus after the canopy has been removed. Iron pan development is common and may give rise to surface-water gleying. Within the cold, wet uplands, some subalpine podzols occur.

The vegetation of the lower ground is dominated by arable communities with wet hollows either drained or supporting rush pasture and sedge mire. Natural Scots pine forests and plantations are locally extensive. Some acid bent–fescue grassland (part of *Achilleo-Festucetum tenuifoliae*) is present on the higher ground which is dominated characteristically by Atlantic or boreal heather moors (Carici binervis-Ericetum cinereae or Vaccinio-Ericetum cinereae). Birchwood is confined mainly to sites of previous cultivation or to sheltered valleys in the uplands. Wet hollows at the latter elevations generally support blanket bogs (Erico-Sphagnetum papillosi).

The soils of this association are maintained both as grassland and moorland or, in the warm, moderately dry lowlands, are cultivated for grain and root crops. The main restrictions on land use are coarse texture, stoniness and an inherently low fertility, the first limitation being associated with problems of windblow and droughtiness. Depending upon the nature of the deposition of the parent material, slope may also influence land management. Moderate to high deficiencies of cobalt and copper are common where there is little argillaceous material in the profile.

Map unit 96   This is a minor unit covering only 14 square kilometres (0.6 per cent of the associations). It occurs within the middle reaches of the Dee valley and the Cawdor oakwood, Nairnshire and is dominated by brown forest soils.

The high proportion of brown forest soils has resulted probably from both the extensive settlement by Bronze-age man accompanied by the persistence of birch cover and from the continuance of the eastern highland oakwood (Trientali-

Betuletum pendulae). The moder surface horizons developed in these circumstances tend to persist even after the invasion by heath vegetation. A common feature of soils of the Dinnet Association is an intensely indurated subsoil horizon.

Map unit 97 By far the largest of the Corby units, it occupies 941 square kilometres (50 per cent of the associations), both within the lower peneplain areas and the coastal lowlands. The latter include wide riverine terraces, for example, those of the Rivers Dee, Don and Ythan, and coastal outwash deposits such as those at Cruden Bay and Belhelvie. Other areas are located within Strathmore along the valleys of the Rivers Tay and Isla, and on the raised beaches and coastal lowlands around the Moray Firth (Plate 9).

Plate 9. *The lower Spey valley and associated raised beaches north of Fochabers. Undifferentiated alluvial soils of map unit 1 (mainly sands and gravels) form the lower river terraces (Class 6.2); they are heavily afforested. The braided nature of the river is evidenced by the white unvegetated areas. On either side, the upper terraces and raised beaches comprise humus-iron podzols of map unit 97 (Corby and Boyndie Associations, Class 3.2). The area is mainly arable and associated with barley production. Droughtiness and coarse textures are the principal limitations. Along the seaward margin there is a narrow fringe of undifferentiated raised beach deposits of map unit 420 (Nigg Association, Classes 5.3 and 6.2); the soils are largely noncalcareous regosols formed in shingle.* Aerofilms.

76

The unit is predominantly agricultural with free-draining, cultivated podzols extensive and noncalcareous and peaty gleys restricted to small localized hollows and depressions. Where the latter soils have been drained and cultivated, they support arable crops, otherwise their vegetation is dominated by rush pasture. In freely drained and cultivated soils, the thickness of the A horizon is generally about 25 centimetres though thicknesses of 40 centimetres are not infrequent; occasionally the horizon exceeds 60 centimetres. Textures range from loamy sand to fine sandy loam. Beneath the plough layer, there is no physical impedance to root development, except for an occasional dark brown, cemented, humus-iron B horizon, the 'Moray Pan'. The natural profile is normally a humus-iron podzol, although iron podzols and humus podzols occur within the Moray Firth area.

With the favourable climatic conditions, barley is widespread though wheat acreage is low due to the coarse texture of the soils. Good drainage conditions and general ease of cultivation tend to be offset by the low moisture- and nutrient-retention capacities of the soil. In dry years, yields are substantially reduced. Continuous cereal cropping has led to increased wind erosion problems along sections of the Moray Firth. Because of these limitations, the unit is mainly capable of producing only a moderate range of crops. Where thicker and anthropogenic topsoils exist, such problems are less and a wider range of crops is possible. The few small areas of poorly or very poorly drained soils are capable of producing a narrow range of crops when drained, otherwise they are restricted to rough grazings. There are minor areas of broadleaved woodlands, mainly eastern highland oak and birchwood (Trientali-Betuletum pendulae), southern oakwood (Galio saxatilis-Quercetum) and Atlantic oakwood and western highland birchwood (Blechno-Quercetum).

Map unit 98   This unit accommodates the soils of some river valleys where areas of alluvial soils and those of the adjacent fluvioglacial terraces are too small to map separately. Occupying 471 square kilometres (25 per cent of the associations), the unit is very widely distributed and is subject to a variable climate. All the alluvial soils have been included as undifferentiated alluvium. On the level fluvioglacial terraces, humus-iron podzols are predominant, most of them being cultivated. A high proportion of commonly adjacent mounds is planted with coniferous plantations or has natural oak and birchwoods. These latter deciduous woodlands are the same as those listed for *map unit 97*. Peaty podzols are of minor extent and are restricted mainly to the climate region of fairly warm, wet foothills. With the exception of a few wet flushes, nearly all the soils of the river terraces are free-draining.

A variety of land use occurs within this unit. Those freely and imperfectly drained alluvial soils which can be worked throughout the year and which have a high inherent fertility, are capable of producing a moderate range of crops. On the lower-lying sites where soil drainage tends to be poor or very poor and where there is an associated risk of flooding, the problems of workability restrict cropping and hinder grassland reclamation. The textural limitations of those soils developed on fluvioglacial material are outlined above. In the cool wet foothill climate regions, wetness is the main limitation but strong slopes are also restrictive.

Map unit 100   Though not an extensive unit, covering 214 square kilometres (11.5 per cent of the associations), its distribution is widespread but restricted to major river valleys and to some areas of glacial outwash plains. The largest area occurs at the Flemington Kames, south-west from Nairn.

Both the soils and topography are distinctive features, the latter in the form of moderately steep or steep-sided kame and esker complexes. The dominant soils are humus-iron podzols and iron podzols, and are largely uncultivated. Minor areas of peaty and humic gleys, peat and alluvial soils also occur. Under semi-natural conditions, Atlantic and boreal heather moors (Carici binervis-Ericetum cinereae and Vaccinio-Ericetum cinereae) are characteristic. Where intensive burning and grazing has occurred, they are replaced by acid bent–fescue grass-land (part of Achilleo-Festucetum tenuifoliae).

The gradient, pattern and frequency of slopes all exert considerable influence on the land use of the unit. On subdued mounds, there is some potential for grassland reclamation; otherwise, steepness of slope inhibits such reclamation.

Map unit 101   Covering 229 square kilometres (12 per cent of the associations), it occurs extensively throughout Eastern Scotland (Sheet 5) within the climate region of fairly warm, wet lowlands and foothills. Major areas are located on Dava Moor, around Lochindorb, within Abernethy and Rothiemurchus Forests and in the general vicinity of Lochs Laggan and Rannoch.

Though the topography is similar to that of the previous unit, the soils differ in that most have a peaty surface horizon and are mainly peaty podzols, with or without an iron pan. On steep-sided mounds, especially where forested, humus-iron podzols exist. Unlike the typical peaty surface horizons associated with the peaty podzol subgroup, the surface organic horizons of soils on sites below 300 metres consist often of compressed L and F horizons which seldom exceed 15 centimetres in thickness. These soils are considered to be transitional between humus-iron and peaty podzols. On flattish or concave sites, strongly gleyed Ag or A/Bg horizons are found commonly above the iron pan and the associated peaty horizon may be up to 40 centimetres thick. Most of the depressions are infilled with peat 2 to 3 metres thick. Limited areas of peaty gleys and alluvial soils are also present.

The peaty surface horizon is an additional limitation to those already discussed for *map unit 100*. In some instances, improved grassland has reverted to poor quality *Juncus* pasture due to inadequate fertilization, or insufficient drainage and maintenance, or both. Much of this unit produces only rough grazing of low value.

Moorland communities, mainly Atlantic and boreal heather moors (Caricie binervis-Ericetum cinereae and Vaccinio-Ericetum cinereae) dominate the peaty podzols, whereas blanket bogs (Erico-Sphagnetum papillosi) typify the peat-filled depressions. Although Scots pine (*Pinus sylvestris*) is the principal tree species, there are a few limited areas of broadleaved woodlands, mainly eastern highland oak and birchwood (Trientali-Betuletum pendulae) and Atlantic oakwood and eastern highland birchwood (Blechno-Quercetum). These deciduous communities are generally located in sheltered valleys.

THE CORRIEBRECK ASSOCIATION

(Map unit 107)

The soils of map unit 107 belong to the Corriebreck Association and are formed from stony drifts derived from basic, ultrabasic and also from acid rocks in local areas. On hill tops and convex slopes, the drift is a brown sandy loam and basic or ultrabasic rock may underlie it within a depth of 1 metre. A sandy clay loam till is usual on lower and concave slopes.

The unit is limited, occurring on undulating lowlands and hills in Glenkindie and in the Nochty valley near Strathdon. It covers approximately 2 square kilometres, less than 0.1 per cent of the land area, and spans an altitude from 275 to 450 metres. Slopes are usually less than 15 degrees. Ultrabasic rock outcrops occur locally on summits, but otherwise the land is non-rocky and non-bouldery.

The average annual rainfall is in the range 1000 to 1100 millimetres. The climatic regions are cool, wet foothill and upland with some cold, wet upland.

The soils of the unit are freely or imperfectly drained brown magnesian soils with some magnesian gleys on the lower concave slopes. Surface horizons tend to be dark-coloured and some are organic. The ratio of exchangeable magnesium to calcium is usually at least 2 to 1 in the subsoil. This contrasts with the ratios of 1 to 5 which are typical of non-magnesian soils, that is with calcium substantially exceeding magnesium. The base saturation of the B and C horizons is in the range 50 to 100 per cent. Although some of the land in the valley floors and lower slopes is cultivated and produces a narrow range of crops based primarily on long ley grassland, most is used for rough grazing and coniferous plantations. Much of the uncultivated land is moderately suited for grassland reclamation.

Plant communities include herb-rich bent–fescue grassland (part of Achilleo-Festucetum tenuifoliae), herb-rich boreal heather moor (part of Vaccinio-Ericetum cinereae) and boreal juniper scrub (Trientali-Juniperetum communis); sedge mires occur on the gleyed soils.

THE COUNTESSWELLS/DALBEATTIE/PRIESTLAW ASSOCIATIONS
(Map units 115–119, 122, 123, 125–129, 131, 134–137)

Only the Countesswells Association is found in Eastern Scotland (Sheet 5). It covers 2253 square kilometres or 9.2 per cent of the land area. With a range of altitude from almost sea level to nearly 1300 metres and a considerable variation in climate and topography, seventeen map units have been used to describe the association which is the third largest within the area.

The soils are developed on material derived from granite, granitic gneiss and other associated granitic rocks, and range from a shallow stony, gritty drift with a coarse sandy loam texture to a loam or sandy clay loam till. The latter is characteristic of lower slopes and depressions which are normally poorly drained. In general, the shallow drift is confined to the upper slopes and foothills. In limited locations where the till or drift is thin or absent, the parent material may consist of deeply weathered granite. At higher altitudes above 800 metres, the parent material is almost entirely locally derived and frost-shattered, imparting a high content of coarse sand and fine gravel to the profile. The parent material also includes moraines and colluvium. The moraines are coarse textured with a high content of subrounded and rounded stones, but the characteristic bedding or stratification of the fluvioglacial deposits is absent. Two notable features of most of the granite soils are the high content of stones, especially subrounded boulders, and the intensive induration.

The granites vary in colour, the pink granites giving rise to a pink drift whereas the grey granites produce a yellowish brown drift. Quartz and feldspar fragments are often conspicuous in these drifts.

Apart from isolated granitic intrusions, for example, at Dufftown, New Pitsligo and Beinn Dearg in the Forest of Atholl, the association is confined to two zones. The eastern zone, which starts in the Cairngorm Mountains, includes Lochnagar, Cromar, Forest of Birse and Hill of Fare and terminates with the Aberdeen granite. The Rannoch Moor granite, the dispersed intrusions of the Monadhliath Mountains, and the Foyers, Moy and Ferness granites form the other zone. The eastern zone is the larger and runs almost continuously from the Cairngorm Mountains to Aberdeen.

Most of the commonly occurring Scottish major soil subgroups are represented in the association. Brown forest soils with free and imperfect drainage are limited and found within only two map units, the undulating lowland *map unit 115* and the steep colluvial *map unit 128*. In the latter, continual rejuvenation of the surface horizons by colluviation inhibits podzolization. Noncalcareous gleys, peaty gleys and humus-iron podzols are common in the lowlands and foothills with peaty podzols at higher elevations. Subalpine soils are widespread above 600 metres and alpine soils above 700 metres. On rugged landforms, rankers are extensive.

Nearly all the arable land is capable of producing a moderate range of crops with a few areas capable of a wider range. Major limitations to land use are climate, wetness, topography and soils. These factors become more severe in the foothills where land use is restricted to a limited arable rotation or long ley grassland. The upper slopes above 450 metres are generally limited to rough grazings with moderate or low relative grazing values. Land between the present arable limits and the upper hill grazings is moderately suited to reclamation and use as improved grassland. Slopes below 550 metres have a considerable potential for afforestation.

Map unit 115   This map unit covers 820 square kilometres, (36 per cent of the association) and is dominant in the undulating north-eastern lowlands and foothills of the River Don and River Dee catchments.

Throughout the area, the soils are developed on a till, or shallow drift, rich in quartz, orthoclase feldspar and biotite. Although some of the till and drift is yellowish brown, the pink type is more widespread. On well-drained areas with convex relief, the parent material is the shallow drift, seldom more than 1.5 metres thick, having a coarse sandy loam texture with many stones and boulders. In several areas on hills and local prominences, the drift is very shallow, less than 0.5 metres, and the shattered granite itself imparts a high content of coarse sand and fine gravel to the drift. In localities where the topography is flat or concave and drainage is poor, the colour of the parent material is modified by gleying, and its texture is usually finer. In these situations, the parent material is a till which varies from 1.5 to over 7 metres thick with a loam or sandy clay loam texture. Principal soils are humus-iron podzols with some brown forest soils and local areas of peat and gleys.

The map unit lies generally below 350 metres and has an average annual rainfall from 750 to 1250 millimetres. Climatically, the unit is associated with the regions of fairly warm, moderately dry lowlands and foothills. Hills and undulating lowlands with gentle and strong, non-rocky slopes are the landforms.

Most of the area is cultivated or has been previously cultivated and is capable of producing a moderate range of crops; one or two minor areas are capable of a wider range. In the foothill areas where there are moderately severe limitations imposed by climate, wetness, topography and soils, land use is restricted to a limited arable rotation or long ley grassland. Some high areas with natural vegetation are suitable for reclamation to grassland.

Plant communities include permanent and ley pastures (Lolio-Cynosuretum) with arable crops on cultivated land. On uncultivated areas, dry boreal heather moor (part of Vaccinio-Ericetum cinereae) is associated with the humus-iron podzols. Locally, dry Atlantic heather moor (part of Carici binervis-Ericetum cinereae) is present.

Map unit 116   Though not an extensive unit, covering 121 square kilometres or 5 per cent of the association, the unit has a very widespread distribution. As with *map unit 115,* map unit *116* occurs mainly throughout the undulating lowland and foothill areas of the River Dee and River Don catchments. The slopes are non-rocky, gentle and strong.

Component soils are noncalcareous and peaty gleys with some humic gleys and peat. They are developed on parent materials similar to those of *map unit 115.* Occupying depressional sites in many instances, the parent material is usually a stony gritty loam or sandy clay loam till, several metres thick. With poor or very poor soils drainage, the colour of the till is often modified by gleying.

The map unit lies generally below 350 metres and has an average annual rainfall from 750 to 1250 millimetres. Climatically, the unit is associated with the regions of fairly warm, moderately dry lowlands and foothills or cool, wet foothills.

In many of the lowland sites, which are cultivated or have been previously cultivated, long-term farming has improved the naturally low fertility of the soils. Some of the arable land produces a moderate range of crops, but generally only limited arable rotations with long ley pastures are possible. Where more severe climatic, topographical or wetness restrictions prevail, cultivation is hindered and land use is limited to permanent grassland.

Plant communities include permanent and ley pastures (Lolio-Cynosuretum) with arable crops on cultivated land. Areas of very poor drainage have soft rush pasture (the *Ranunculus repens–Juncus effusus* Community) and sedge mires. On uncultivated areas, bog heather moors (Narthecio-Ericetum tetralicis) and blanket bogs (Erico-Sphagnetum papillosi) dominate.

Map unit 117   Located in the drier, eastern hills, this map unit covers 292 square kilometres or 13 per cent of the association. The main distribution is in the foothills of south Deeside. Other minor areas include Bennachie and the Hill of Fare.

The soils are peaty podzols with some humus-iron podzols and gleys. A stony gritty sandy loam drift forms the soil parent material which, on strong slopes, is shallow. On local prominences where the drift is very shallow, less than 0.5 metres, the weathering of the shattered granite contributes a high content of coarse sand and fine gravel.

Climatically, the map unit is associated with the cool, wet foothills and uplands region as well as the cold wet upland region. Both altitude and average annual rainfall ranges are wide, 90 to 600 metres and 850 to 1400 millimetres respectively. Hills and valley sides with gentle to steep, non-rocky slopes are the principal landforms.

Cultivation is less practicable on this map unit due to the ground pattern and the increased thickness of the organic surface horizon. Over most of the area there is a substantial capability for grassland reclamation, but as the climatic and topographical limitations increase in severity, land use is restricted to rough grazing; the grazings are of low value, occasionally moderate.

Plant communities are boreal and Atlantic heather moors (Vaccinio-Ericetum cinereae and Carici binervis-Ericetum cinereae respectively), bog heather moors

(Narthecio-Ericetum tetralicis) and some native pinewood (Pinetum scoticae). Areas which have been reclaimed to grassland have a permanent pasture (Lolio-Cynosuretum) community.

Map unit 118   Occupying only 4 per cent of the association, 87 square kilometres, this map unit is limited in distribution but is locally widespread on the granite around the River Gairn, Glen Fenzie, Glen Livet and south-west of Dufftown. Other areas have also been mapped in the hills to the north of Moy, south-west of Tomatin and near Lochnagar. With the increasing average altitude and rainfall, the organic surface horizons are thicker than those of the previous unit, thereby increasing the percentage of peat and peaty soils; the latter are mainly peaty podzols with some peaty gleys.

The soil parent material is a stony gritty sandy loam drift, often shallow, with a high content of coarse sand and fine gravel derived from the underlying shattered granite. The landform is one of undulating uplands and hills with gentle and strong, non- and slightly rocky slopes. With an altitude range from 250 to over 650 metres and an average annual rainfall between 1000 and 1400 millimetres, climatically the unit belongs to the cool wet foothills and uplands region and the cold, wet uplands region.

Limitations to agriculture are poor or very poor surface drainage, a thick organic horizon, climate and topographical pattern. Generally, the land use is restricted to rough grazing which has a low relative grazing value. In a few areas, where the organic surface horizon is not too thick, the map unit may be well suited to grassland reclamation.

Plant communities include boreal and Atlantic heather moors (Vaccinio-Ericetum cinereae and Carici binervis-Ericetum cinereae respectively), bog heather moors (Narthecio-Ericetum tetralicis) and lowland and upland blanket bogs (parts of Erico-Sphagnetum papillosi).

Map unit 119   Few areas of map unit 119 are found in Eastern Scotland (Sheet 5) and they occupy only 24 square kilometres (1 per cent of the association). Map unit 119 is similar to, but wetter than, *map unit 118* and its distribution is mainly in wetter, western areas. Principal locations are Rannoch Moor and upper Glen Glass.

Dominant soils are peaty gleys and peat with some peaty podzols. The soil parent material is usually a shallow sandy loam drift, often containing a moderately high percentage of locally derived stone fragments. In some of the larger depressional and bench-like areas, a thicker, stony gritty sandy loam, locally loam, till is present.

Undulating uplands and hills with gentle and strong, non- to slightly rocky slopes are the landforms. With an altitude range from 300 to 500 metres and an average annual rainfall of 1200 to 1600 millimetres, the unit is associated climatically with the cool, wet foothills and uplands region.

Because of the severe restrictions of wetness, climate, rockiness and the organic surface horizons, limited areas are only marginally suited to grassland reclamation. Most of the unit is restricted, however, to rough grazing; these grazings are of low value.

Common plant communities include bog heather moors (Narthecio-Ericetum tetralicis), lowland and upland blanket bogs (parts of Erico-Sphagnetum papillosi) and on slightly drier sites, moist boreal heather moor (part of Vaccinio-Ericetum cinereae).

Map unit 122   Occupying less than 1 square kilometre (less than 1 per cent of the association) this unit is found in the Gaur valley, near the western end of Loch Rannoch. Most of the soils have been improved by previous cultivation and intense grazing has altered much of the natural vegetation. Dominant soils are humus-iron podzols, although beneath the bracken, humus-iron podzols transitional to brown forest soils are present. Some peaty gleys and peat occur in depressions.

The soil parent material is a stony, often very stony, gritty, sandy loam or loamy sand moraine. In the depressions, the parent material may be a finer-textured drift, often greyer in colour due to gleying. Hummocky valley moraine, frequently bouldery, is the dominant landform. Some of the hummocks are rock-cored. With an altitude of approximately 300 to 350 metres and an average annual rainfall of 1300 millimetres, the map unit is in the climate region of cool, wet foothills and uplands.

The major limitations to land use are a complex slope pattern, surface boulders and climate.

Map unit 123   Covering 125 square kilometres, this unit comprises nearly 6 per cent of the association. Although a few minor areas exist, by far the most extensive area is associated with Rannoch Moor.

Principal soils are peaty podzols, peat and peaty gleys. The parent material is a stony, often very stony, gritty, sandy loam or loamy sand moraine. Although most of the moraines contain many rounded stones, some have much locally derived subangular materials. Occasionally the mounds are rock-cored.

The altitude range is 300 to 700 metres and the map unit has an average annual rainfall of the order 1200 to 2200 millimetres. Climatically, the map unit is associated with the regions of cool, wet foothills and uplands or the cold, wet uplands. The landform is one of hummocky valley and slope moraines, often with a bouldery surface.

Major limitations to land use are the poor drainage of the organic surface horizons and the complex topographical pattern. As a consequence, most of these areas are restricted to rough grazing which is of a low relative grazing value.

Plant communities include moist boreal and Atlantic heather moors (parts of Vaccinio-Ericetum cinereae and Carici binervis-Ericetum cinereae respectively), flying bent–bracken grassland (the *Molinia caerulea–Pteridium aquilinum* Community), bog heather moors (Narthecio-Ericetum tetralicis) and blanket bogs (Erico-Sphagnetum papillosi).

Map unit 125   Covering only 4 square kilometres (less than 1 per cent of the association), this is one of the least extensive units in the Countesswells Association. There are only two locations, one on the steep, colluvial slopes above Loch Ericht and the other in Glen Tilt. Component soils are humus-iron podzols and brown forest soils, with some noncalcareous gleys and podzolic rankers.

The soil parent material is a stony, gritty sandy loam drift of variable thickness, often including many locally derived subangular and angular rock fragments. Associated with the base of the steeper slopes, there is a finer-textured, stony colluvium in which brown forest soils are developed, the constant colluviation of the upper horizons inhibiting development of a peaty or podzolic soil.

With an average annual rainfall of 1600 millimetres and an altitude of 250 to 700 metres, the map unit is climatically in the cold, wet uplands region. The landform is one of valley sides with strong to very steep, moderately rocky and bouldery slopes.

Limitations to land use are climate, rockiness and slope. These restrictions are very severe and land use is limited to rough grazing; the grazings are of a moderate value.

Plant communities include acid bent–fescue grassland (part of Achilleo-Festucetum tenuifoliae), soft rush pasture (the *Ranunculus repens–Juncus effusus* Community) and sedge mires, and birchwoods (part of Blechno-Quercetum).

Map unit 126   This map unit occupies an area of 58 square kilometres or approximately 3 per cent of the association. Although limited in extent, the map unit has a widespread distribution between Loch Moy and Loch Rannoch. Dominant soils are peaty podzols and humus-iron podzols with some peaty gleys and rankers.

The parent material is a stony, sandy loam drift. This may also include variable quantities of locally derived subangular and angular rock fragments. Hills and valley sides with strong to very steep, moderately rocky slopes are the dominant landforms. With an altitude range from 200 to over 750 metres and an average annual rainfall from 1000 to over 1600 millimetres, the map unit is mainly in the climate region of cool wet foothills and uplands although, at its highest elevations, it is within the cold wet uplands.

The major land use restrictions are rockiness, organic surface horizons and slope complexity. Because these limitations are usually very severe, land use is largely confined to rough grazing; the grazings form a mosaic of low and moderate grazing values. Where the organic surface horizon is not too thick and a reasonable percentage of non-rocky ground is available, then grassland reclamation is marginally possible.

Plant communities include boreal heather moors (Vaccinio-Ericetum cinereae), flying bent grassland (part of Junco squarrosi–Festucetum tenuifoliae) and boreal juniper scrub (Trientali-Juniperetum communis).

Map unit 127   The unit covers an area of 57 square kilometres or 3 per cent of the association and is often found associated with *map units 119, 126* and *129*, in the wetter and higher western areas, for example, the Corrieyairack Forest and Rannoch Moor. Dominant soils are peaty gleys and peat with some peaty podzols and peaty rankers.

The map unit has a wide range of both altitude and average annual rainfall, 300 to over 800 metres and 1200 to over 2200 millimetres respectively. Climatically, the map unit is mainly in the regions of cool, wet foothills and uplands, although at higher elevations, for example, above 600 metres, the map unit is within the cold wet uplands region.

A shallow, stony, gritty sandy loam drift forms the soil parent material, often with a variable proportion of locally derived, fragmented rock. Hills and undulating lowlands, with gentle and strong moderately rocky slopes, are the landforms. Land use limitations are surface wetness, climate, rockiness, slope pattern and thick organic surface horizons. These restrictions are usually very severe and only rough grazing is possible; the relative grazing values are low. Areas may be afforested but windthrow may be a problem.

Plant communities include moist boreal and Atlantic heather moors (parts of Vaccinio-Ericetum cinereae and Carici binervis-Ericetum cinereae respectively), bog heather moors (Narthecio-Ericetum tetralicis), and lowland and upland blanket bogs (parts of Erico-Sphagnetum papillosi).

Map unit 128   As one of the less extensive map units in Eastern Scotland (Sheet 5), it occupies only 3 square kilometres (less than 1 per cent of the association) at one location near the northern end of Loch Ness. Principal soils are brown rankers and brown forest soils with some humus-iron podzols. The parent material on which the soils are developed is a stony, sandy loam colluvium with a very high percentage of locally derived rock fragments. The unit is restricted to a very steep, south-east-facing slope covered with active or stabilized scree mostly below 400 metres. Climatically, it is associated with the regions of fairly warm, moderately dry lowlands and foothills, and the cool wet foothills and uplands on the upper slopes. The topography limits land use mainly to rough grazings; the grazings are of moderate value.

Plant communites include acid bent–fescue grassland (part of Achilleo-Festucetum tenuifoliae), dry Atlantic heather moor (part of Carici binervis-Ericetum cinereae), Atlantic oakwood and western highland birchwood (Blechno-Quercetum) and western ash-oakwood (Primulo-Quercetum).

Map unit 129   This map unit is similar to *map unit 126* but the landscape is rockier. Covering 47 square kilometres, or approximately 2 per cent of the association, it is normally found with *map units 118* and *126*. Rankers and peaty podzols are the dominant soils, with some humus-iron podzols and peaty gleys.

The parent material is a stony, gritty sandy loam drift of variable thickness, often including much locally derived, subangular and angular fragmented rock. Associated with the base of the steep slopes, there is a finer-textured, stony colluvium. Rugged hills with strong and steep, very rocky slopes with some scree are the landforms.

With an altitude range of 50 to 600 metres and an average annual rainfall from 900 to 1600 millimetres, the map unit occurs in the climatic regions of cool, wet foothills and uplands or the cold, wet uplands.

Limitations to land use are rockiness, soil, complex slope pattern and thick organic surface horizons. These restrictions are very severe and limit agricultural use to rough grazing; grazing values are mainly low. Because of the rocky nature of the map unit, afforestation is only possible by hand planting and the extraction of the timber crop requires considerable care.

Plant communities include boreal heather moors (Vaccinio-Ericetum cinereae), bog heather moors (Narthecio-Ericetum tetralicis) and heath rush–fescue grasslands (Junco squarrosi-Festucetum tenuifoliae).

Map unit 131   This map unit, occupying only 9 square kilometres or less than 1 per cent of the association, is one normally associated with the wet west coast of Scotland. It is found in the high rainfall area of Loch Ossian where it occurs on the very steep, wet, valley sides. The parent material is a stony, gritty, sandy loam drift of variable thickness and usually contains a very high percentage of locally derived, subangular or angular fragmented rock. Associated with the base of the steeper slopes, there is a finer-textured, stony colluvium.

Dominant soils are peaty gleys, peaty podzols and peaty rankers and the associated landform is one of hill sides with steep and very steep, moderately or very rocky slopes. The altitude range is 350 to over 600 metres with an average annual rainfall of 1600 to over 2000 millimetres. Climatically, the map unit is in the cold wet uplands region.

Limitations to land use are complex steep slopes, wetness, climate and rockiness. These limitations are very severe but the colluvial nature of the parent material provides for continual enrichment of the soil profile, as does the considerable flushing within this map unit. Rough grazing is the principal land use; relative grazing values are moderate.

Plant communities include heath rush–fescue grasslands (Junco squarrosi-Festucetum tenuifoliae) and bog heather moors (Narthecio-Ericetum tetralicis).

Map unit 134 This unit covers 70 square kilometres or 3 per cent of the association. It is fairly widespread throughout the association on mountains with gentle to very steep, non- to very rocky slopes above 650 metres. The soil parent material is a shallow, very stony, sandy loam drift, sometimes very gritty because the weathering of locally derived granite rock debris imparts a high content of coarse sand and fine gravel to the profile. In limited locations, where the drift is absent, the parent material may consist of deeply weathered granite. Dominant soils are subalpine, mainly podzols, with some peat and rankers.

The altitude of the unit is usually between 650 and 800 metres with an average annual rainfall above 1600 millimetres. Climatically, the map unit is mostly in the cold wet uplands region. Some of the higher areas with less favourable conditions are designated as very cold, wet mountains.

Land use limitations are topography, rockiness, soil and climate. These limitations are very severe and only rough grazing is possible; grazings are mainly of low value. Some small areas of snow-bed vegetation afford moderate grazing values as do peat flushes, for a limited period of the year.

Plant communities on the drier subalpine soils include alpine azalea–lichen heath (Alectorio-Callunetum vulgaris) and lichen-rich boreal heather moor (part of Vaccinio-Ericetum cinereae) whereas on the peat, mountain blanket bog (Rhytidiadelpho-Sphagnetum fusci) dominates.

Map unit 135 Occupying 7 per cent of the association, this unit covers 153 square kilometres. Related to eroding blanket peat (*map unit 4e*), it is widespread within higher levels of the association. The main areas are in the eastern Cairngorms Mountains and around the headwaters of the River Findhorn in the Monadhliath Mountains. Component soils are peat and subalpine soils with some alpine soils; the mineral soils are mainly podzolic. A frost-shattered rock debris, or a thin, stony, sandy loam drift often with a stone pavement on or at the surface, forms the parent material.

The altitude range of the map unit is 600 to over 800 metres and the average annual rainfall is over 1600 millimetres. Climatically, the unit is associated with the cold wet uplands and the very cold, wet mountain regions. Mountains with gentle and strong, non- to moderately rocky slopes are the landforms.

Extremely severe restrictions in climate, exposure, soil, wetness and topography limit the land use to very poor seasonal grazings in the eastern Cairngorm Mountains. At lower levels, for example, near Ballochbuie, the limitations are less severe and the rough grazings have low values.

Plant communities on the peat include mountain blanket bog (Rhytidiadelpho-Sphagnetum fusci) with alpine azalea–lichen heath (Alectorio-Callunetum vulgaris) and stiff sedge–fescue grasslands (the *Carex bigelowii–Festuca vivipara* Association) on the drier and podzolic subalpine and alpine soils.

Map unit 136 This is the most extensive of the mountain map units, covering 211 square kilometres or 9 per cent of the association. The map unit occurs regularly on level summits above 800 metres where the soils are mainly alpine, with minor areas of rankers, peat, and subalpine soils. Most of the alpine and subalpine soils are podzolic and freely drained.

Normally, a shallow frost-shattered and weathered mountain-top detritus forms the parent material. Locally, a thin veneer may overlie completely weathered rock. As with *map unit 135*, extensive areas are located in the eastern Cairngorm Mountains. With an average annual rainfall in excess of 1800 millimetres and extremely severe winters, the map unit is associated with the very cold mountain climate region. The landform is one of mountain or hill summits with gentle and strong, non- to very rocky slopes, often bouldery.

Land use limitations are extremely or very severe, and include topography, climate, erosion and soil. Only very poor rough grazings of a seasonal nature are available where the vegetation is sparse, for example, Ben Avon, Mount Keen and Lochnagar. At lower levels where conditions are less severe, there are rough grazings mainly of low value; snow-bed vegetation often provides local areas of moderate grazing values.

Plant communities are alpine azalea–lichen heath (Alectorio-Callunetum vulgaris), stiff sedge–fescue grassland (the *Carex bigelowii–Festuca vivipara* Association) and three-leaved rush heath (Cladonio-Juncetum trifidi).

Map unit 137   This mountain unit covers 168 square kilometres, 7 per cent of the association. A large proportion of the areas mapped consists of the bare rock pavements and boulder-fields of the Cairngorm Mountains, particularly on the summits of Einich Cairn, Braeriach, Cairn Toul and Ben Macdhui (Plate 10). There are two characteristic landforms associated with this map unit; both are very rocky. A very steep form is confined to mountains with well-developed corries, for example, Braeriach, and to rocky, very steep and complex scree

Plate 10. *The summit plateau of Braeriach (1296 metres). Exposure is extreme and there are less than 300 day-degrees centigrade of accumulated temperature. The surface is covered by granite boulders and weathered granitic detritus. Apart from minor patches of alpine soils, rankers and lithosols of map unit 137 (Countesswells Association, Class 7) dominate the area. Most soils are subject to frost heave. Vegetation cover is sparse and characterized by* Juncus trifidus *and* Racomitrium lanuginosum. *In minor snow-beds where cover is usually complete, the dominants are* Nardus stricta, Carex bigelowii *and* Deschampsia flexuosa.

slopes. This type of slope occurs also in the deeply incised valley of the upper Findhorn. A horizontal form relates to the rock pavement and boulder-fields of the mountain summits.

The soil parent material on the mountain tops is locally derived frost shattered and weathered rock debris whereas that of the steep slopes is scree and a gritty sandy loam colluvium. Component soils are rankers and lithosols; some alpine soils occur in limited areas on the summits.

Because of the physical nature of the map unit, the altitude range is wide, from the upper Findhorn valley at 400 metres to over 1300 metres in the Cairngorm Mountains, where the average annual rainfall is in excess of 2000 millimetres. Climatically, the summit areas are very exposed with very severe winters and belong to the very cold wet mountain region. On the high levels, land use limitations are extremely severe and only limited seasonal grazings are available, restricted mainly to areas of snow-bed vegetation. Moderate grazing values are often associated with the vegetation on the sheltered lower valley sides.

On the highest plateaux in the Cairngorm Mountains, the vegetation is sparse and the dominant plant community is the three-leaved rush heath (Cladonio-Juncetum trifidi). Elsewhere other plant communities include blaeberry and bog whortleberry heaths (the *Rhytidiadelphus loreus–Vaccinium myrtillus* and the *Racomitrium lanuginosum–Vaccinium uliginosum* Community respectively), alpine azalea–lichen heath (Alectorio-Callunetum vulgaris) and alpine clubmoss snow-bed (the *Lycopodium alpinum–Nardus stricta* Community).

THE CRAIGELLACHIE/POLFADEN ASSOCIATIONS

(Map unit 140)

The soils of both the Craigellachie and Polfaden Associations are developed mainly on fluvioglacial silts. They cover 12 square kilometres (less than 0.1 per cent of the land area) and the two associations are geographically distinct. For the 1:250 000 map, however, they have been grouped together; there is only one map unit and its description has been included with that of the association.

Occasionally, the parent material of the Craigellachie Association also includes thin bands of sands and gravels as well as rare patches of till of the Aberlour Association. Silty clay and clay are major components south of Craigellachie. Sands and gravels mainly in the form of a thin mantle, though occasionally as thin seams or lenses, are minor components of the parent material of the Polfaden Association; it is usually distinctly bedded, glaciolacustrine silt or clay.

The Craigellachie Association occurs mainly between Charlestown of Aberlour and Craigellachie on a non-rocky gently sloping terrace, 100 to 200 metres above the River Spey. There is also a small area near Longmorn, south of Elgin. The climate region is that of fairly warm moderately dry lowlands and foothills and the average annual rainfall is below 900 millimetres.

The Polfaden Association occurs in one small area north-east of Inverness on non-rocky, gently undulating land at an altitude of below 50 metres. Other minor areas, not reproducible at the 1:250 000 scale, exist along the coastal belt as far as Nairn. The climate region is that of warm dry lowlands; the average annual rainfall is below 700 millimetres.

Cultivated humus-iron podzols are the main soils of map units 140 which also includes some gleys, mainly noncalcareous. In the Spey valley, the podzols are mostly imperfectly drained. On the raised beaches, however, both freely and imperfectly drained podzols occur, the freely drained often having deep anthro-pogenic topsoils of 40 centimetres or more (plaggen soils).

Most of the land is cultivated and capable of producing a moderate range of crops with the plaggen soils capable of a wide range. Arable and permanent pastures are the principal plant communities, with areas of soft rush pasture (the *Ranunculus repens–Juncus effusus* Community) and sedge mires in local wet hollows.

THE CROMARTY/KINDEACE ASSOCIATIONS

(Map units 144–146)

The parent materials of this soil association grouping are drifts derived from Middle and Upper Old Red Sandstone strata. They comprise a compact reddish brown sandy loam, loam or sandy clay loam till (Cromarty Association) and a stony moderately coarse-textured and partially water-sorted morainic drift overlaying the compact till at depth (Kindeace Association).

The associations, covering 174 square kilometres (0.7 per cent of the land area), are restricted to the undulating, coastal lowlands and foothills around the Moray Firth from Easter Ross to the River Spey; the slopes are gentle and strong. The altitude ranges from sea level to a maximum of 220 metres, south-east of Inverness.

Average annual rainfall is between 550 and 900 millimetres. The climate regions are the warm moderately dry lowlands and the fairly warm moderately dry lowlands and foothills.

The soils are mostly cultivated, the dominant soils being freely or imperfectly drained humus-iron podzols. Poorly drained soils, comprising noncalcareous, humic and peaty gleys and peat, are minor components.

Most of the low ground is in arable agriculture whereas much of the higher ground is afforested. The non-forested foothills are used mainly as rough grazing with some used as grouse-moors.

Map unit 144   This unit comprises humus-iron podzols, with some gleys, and occurs on undulating lowlands with mainly gentle but locally strong slopes. The parent material is mostly a compact, reddish brown loam till, occasionally sandy clay loam; on some low hills in Easter Ross the shallow drift (Kindeace Association) overlies soft red sandstone.

Occupying 40 square kilometres, the unit is found around Nigg in Easter Ross and in the central and northern sections of the Black Isle.

Imperfectly drained podzols are dominant although freely drained podzols occur on shedding sites and steeper slopes. Noncalcareous gleys are of limited extent.

In the podzols, the texture of the topsoils is usually a sandy loam with 12 per cent clay, or more. The B horizons are of a similar texture and are normally indurated. In the higher parts of the Black Isle, such induration is often less than 30 centimetres from the surface; elsewhere it is deeper, often below 50 centimetres. The percentage of clay sometimes increases down the profile to 14–25 per cent in the C horizon, textures ranging from sandy loam to sandy clay loam. Stone content is moderate, sandstones being dominant.

Plant communities are mainly 'arable and permanent pastures (Lolio-Cynosuretum). Soft rush pasture (the *Ranunculus repens–Juncus effusus* Community) and sedge mires occur on the noncalcareous gleys.

Most of the land is cultivated and ranks amongst the most productive in Easter Ross and the Black Isle. The lowland soils are capable of producing a wide or moderate range of crops with high yields whereas the limited areas of soils on the higher slopes are restricted by climate and shallowness to a moderate range with average yields. A few of the wetter and higher areas are used for rough grazing or are afforested.

Map unit 145   This unit consists of humus-iron podzols with some gleys and is found on undulating lowlands and foothills with gentle and strong slopes. The parent material is a partially water-sorted, morainic drift (Kindeace Association) which overlies the till of the Cromarty Association.

The unit covers 115 square kilometres (65 per cent of the associations), its main expression being on the slopes of the foothills above the Moray Firth Lowlands. It includes parts of Drummossie Muir and Darnaway Forest to the south of the Moray Firth; in Easter Ross it forms a belt from north of Alness to beyond Invergordon.

The podzols are mainly imperfectly drained with an indurated B horizon causing an impedance to drainage. Freely drained podzols are restricted to shedding sites; these sites are often ridges orientated south-west to north-east. The topsoil has a stony sandy loam texture and overlies a coarser texture, usually loamy sand, in the B horizon. The underlying C horizon texture is a sandy loam, loam or sandy clay loam. Normally, the soils are stonier than those in *map unit 144,* many of the stones being acid Moine schists. Peaty and noncalcareous gleys occupy hollows and receiving sites. Peaty podzols occur locally, especially at higher levels.

Plant communities are arable and permanent pastures (Lolio-Cynosuretum) with rush pastures and sedge mires associated with the noncalcareous gleys. Uncultivated podzols support boreal and Atlantic heather moors (Vaccinio-Erictum cinereae and Carici binervis-Ericetum cinereae). Bog heather moors (Narthecio-Ericetum tetralicis) dominate the peaty gleys, with lowland and upland blanket bogs (parts of Erico-Sphagnetum papillosi) on the peat.

Much of the land is cultivated and capable of producing a narrow range of crops based on grassland. The main limitations affecting agricultural capability are stoniness and shallowness, the latter due to induration. Substantial areas are afforested with coniferous plantations especially around Darnaway and on Drummossie Muir. Moorland used as rough grazing occurs in Easter Ross and on Drummossie Muir.

Map unit 146   Consisting of noncalcareous gleys and peaty gleys with some humic gleys and peat, this unit occurs in hollows and receiving sites in undulating lowlands and foothills. Slopes are mainly gentle, rarely strong.. The parent materials are variable and include the compact till (Cromarty Association) and the morainic drift (Kindeace Association) together with colluvium.

The unit covers 19 square kilometres (10 per cent of the associations). In Easter Ross, it occupies three small, mostly cultivated areas of noncalcareous gleys whereas, to the south of the Moray Firth, it occurs in three, mainly uncultivated areas of peaty gleys. A large area to the south-west of Forres in afforested.

Wetness and climate are the main limiting factors affecting the agricultural capability.

THE DARLEITH/KIRKTONMOOR ASSOCIATIONS
(Map units 147, 149, 150, 153, 154 and 158)

The soils in this group are developed on drifts derived from basaltic lavas and various basic intrusive rocks. Only the Darleith Association is found in Eastern Scotland (Sheet 5). It covers 191 square kilometres (0.8 per cent of the land area) and occurs on the Sidlaw Hills in Angus, on the basaltic hills and the volcanic ridges of Fife and Kinross and on the Gargunnock Hills of Stirling. First mapped in Ayrshire, the association originally covered soils developed on drifts derived mainly from lavas of Calciferous Sandstone age. Since then, the parent material has been extended to cover drifts derived from a wider range of basic igneous rocks and as a result, a range of textures is found among the derived soils. Fine-textured soils are developed on drifts derived from basaltic lavas or basic intrusive rocks, and coarse-textured soils are derived mainly from dolerites and basic agglomerates.

Six map units of the association have been distinguished in Eastern Scotland (Sheet 5).

Map unit 147   The most extensive unit, map unit 147, comprises freely drained brown forest soils, some imperfectly drained brown forest soils with gleying and poorly drained noncalcareous gleys. It covers 138 square kilometres (70 per cent of the association). The dominant soils in the unit, the freely drained soils, are generally characterized by a loam topsoil with well-developed crumb structure; the texture of the subsoil varies from loam or gritty loam to sandy loam or, sometimes, loamy sand according to the nature of the parent rock. The unit occurs mainly on undulating lowlands with gentle slopes and on some hill ground. Much of the land is under arable cultivation and produces a moderate range of crops. Where shallowness of soil or steepness of slope become limiting factors, the range is narrow and is primarily based on long ley grassland. Other areas support bent–fescue grasslands (Achilleo-Festucetum tenuifoliae) or broadleaved woodland.

Map unit 149   This is the least extensive map unit of the Darleith Association in Eastern Scotland. It occurs in only two locations, both of which are less than 1 square kilometre. One of these units is on the Braes of the Carse, Hallyburton Forest and the other just to the west of Lundie in Angus.

Dominant soils are noncalcareous and humic gleys with some brown forest soils. The topography is one of undulating lowlands with gentle and strong slopes. These slopes are usually non-rocky. Although the fine-textured loam to sandy clay loam topsoils and clay loam subsoils are difficult to manage, much of the land is under cultivation. A narrow to moderate range of crops is obtained, but where wetness is too severe a restriction for cropping, permanent or ley pastures are dominant.

Plant communities include arable and permanent pastures (Lolio-Cynosuretum) on the mineral soils. On the humic gleys, communities are dominated by soft rush pasture (*the Ranunculus repens Juncus effusus* Community) and sedge mires.

Map unit 150   This unit is closely related to *map unit 147* but is moderately rocky and is found often at higher altitudes over a wider range of topography. Occupying 10 per cent of the association, it covers 18 square kilometres and comprises brown forest soils and some brown forest soils with gleying, together with minor areas of brown rankers and noncalcareous gleys. With limitations due to slope and to rock outcrops, land use is restricted mainly to pasture. Some of the area is moderately suited to grassland reclamation whereas other areas provide only rough grazing of moderate grazing value. The limited areas of arable agriculture can grow only a narrow range of crops based primarily on grassland. Plant communities include bent-fescue grasslands (Achilleo-Festucetum tenuifoliae) and common white bent grassland (part of Junco squarrosi-Festucetum tenuifoliae). Crested hair-grass grassland (the *Galium verum–Koeleria cristata* Community) is associated with the rankers.

Map unit 153   Consisting of peaty podzols and minor areas of humus-iron podzols with some small areas of brown forest soils, noncalcareous gleys and peaty gleys, it is a minor unit and occupies 10 square kilometres (5 per cent of the association). Occurring on the Lomond Hills east of Loch Leven and on the Gargannock Hills to the west of Stirling, the unit is slightly rocky and is found on gentle and strong slopes at altitudes from 300 to 500 metres. The vegetation is dominantly common white bent grassland (part of Junco squarrosi-Festucetum tenuifoliae) with minor areas of dry and moist Atlantic heather moors (parts of Carici binervis-Ericetum cinereae).

Land use is restricted and most areas are only moderately suited to grassland reclamation. The main limitations are slope, rockiness and altitude; where these limitations are very severe, the land use is further restricted to rough grazing. Grazing values are moderate.

Map unit 154   This unit comprises peaty podzols, peaty gleys and peat with some rankers, and occurs on hills with gentle and strong slopes. It covers 8 square kilometres (less than 5 per cent of the association).

The vegetation is dry and moist Atlantic heather moors (parts of Carici binervis-Ericetum cinereae) with common white bent grassland (part of Junco squarrosi-Festucetum tenuifoliae) and bog heather moors (Narthecio-Ericetum tetralicis). Slope, altitude and rock outcrop limit land use to improved grassland or rough grazing. Grazing values are low, occasionally moderate.

Map unit 158   Map unit 158 has, as its main component, freely drained brown forest soils; it is closely related to *map unit 147* but is found on hills and valley sides at altitudes from 90 to 300 metres. The soils are slightly rocky and the landform is occasionally terraced with gentle, strong and sometimes steep slopes. The unit, which has been mapped in Fife and in east Perthshire, also includes some brown rankers with small areas of humus-iron podzols and peaty podzols. It covers 17 square kilometres (10 per cent of the association). Much of the brown forest soil area is under permanent pasture with bent–fescue grasslands (Achilleo-Festucetum tenuifoliae) dominant; rankers and other soils support herb-rich Atlantic heather moor (part of Carici binervis-Ericetum cinereae).

Limitations of slope, rockiness and occasionally altitude are the main restrictions on land use. Most areas are moderately suited to grassland reclamation. A few areas where limitations are more severe are restricted to rough grazing. Grazing values are moderate and high.

THE DARVEL ASSOCIATION

(Map unit 163)

The soils of the Darvel Association are developed on sands and gravels derived mainly from Carboniferous igneous and sedimentary rocks with some Old Red Sandstone material. They occur in scattered areas throughout the lowland regions of Fife, Stirling and east Perthshire, covering a total area of 36 square kilometres (0.2 per cent of the land area). Only map unit 163, has been mapped in Eastern Scotland (Sheet 5), occurring on mounds and terraces with gentle and strong slopes. It consists of freely drained brown forest soils and imperfectly drained brown forest soils with gleying together with some humus-iron podzols and noncalcareous gleys.

Most of the unit is cultivated and produces a moderate range of crops with high or average yields. A few areas, where there is a deeper topsoil of imperfect drainage associated with the more level ground, produce a wide range of crops. Coarse texture and stoniness, often combined with the added restriction of moundy topography, are the main limitations.

THE DEECASTLE ASSOCIATION

(Map unit 165)

The Deecastle Association is one of the less extensive soil groupings, occupying only 33 square kilometres (0.1 per cent of the land area) within isolated locations which extend south from Tomintoul and upper Deeside into central and eastern Perthshire. The soils are formed from Dalradian limestone and calc-silicate rocks and their derived drifts. Because the rock is seldom exposed and the drift is often mixed, the association is frequently indicated only by a base-rich vegetation. This vegetation, however, is not specific as it can often extend into surrounding soils which are developed on a totally different parent material but are flushed with enriched drainage water. This enrichment is consequent upon the drainage water crossing adjacent outcropping limestone. Only map unit 165 has been recorded in Eastern Scotland.

The texture of the parent material is dominantly loam, or fine sandy loam, with some sandy loam and loamy sand where the drift is shallow and very stony. Colluvium occurs occasionally on steep slopes.

Most of the association area is occupied by brown forest soils, which differ from similar soils on less basic parent materials by having greater amounts of exchangeable calcium and magnesium in the C horizon and a less brightly coloured B horizon with a chroma usually 4 or less. Where the soil reaction is neutral, brown calcareous soils are distinguished. Rendzinas and brown rankers are found only in small areas. Calcareous gleys are restricted to concave slopes and to other slopes which are flushed.

The association is found mainly in the climate regions of cool, or cold, wet foothills and uplands in the Tomintoul area and upper Deeside. It occurs also within the cool or fairly warm, wet foothills of Perthshire. Species-rich communities have been recorded on all the free-drained soils with rockrose–fescue grassland (the *Galium sterneri–Helictotrichon pratense* Community) and herb-rich bent–fescue grassland (part of Achillio-Festucetum tenuifoliae) being typical examples. At higher elevations, where there is a surface organic build-up and podzol development, the herb-rich boreal heather moor (part of Vaccinio-

Ericetum cinereae) is locally extensive and may often contain boreal juniper scrub (Trientali-Juniperetum communis).

Apart from arable agriculture, the association is utilized for rough grazing. Some of the marginal land was previously under cultivation but climate restricted output to low or moderate yields. Because most of the natural or semi-natural grassy vegetation has a high grazing value, reclamation is not often practised. Apart from the presence of rock at or near the surface, there are few soil limitations. Where elevation coincides with unfavourable site conditions for reclamation and where grazing pressure is less severe, the resultant herb-rich boreal heather moor (part of Vaccinio-Ericetum cinereae) has a moderate grazing value.

Only one map unit has been recognized and it contains all the soils mentioned for this association. Occurring on strongly undulating and hilly land, the unit often has a complex topography of ridges or knolls and associated hollows. The ridges are rock-controlled and may be slightly or moderately rocky. Around Inchrory, in the upper Avon valley, the rock outcrop is associated with substantial areas of scree which are often stabilized. Where environmental conditions such as altitude and slope are favourable, this unit is used for arable agriculture and produces a narrow range of crops. Such land exists around Tomintoul and Blair Atholl. With the exception of isolated areas where mechanized surface treatments to improve the grassland are possible, the rest of the unit provides rough grazing of high value.

THE DOUNE ASSOCIATION

(Map unit 168)

The soils of the Doune Association are developed on fluvioglacial sand and gravel derived from Dalradian acid schists and Lower Old Red Sandstone sediments, mainly hard, red sandstones with some lavas and intrusive rocks, mainly andesites and basalts. Most of the parent material is gravelly though there are frequent sandy, almost gravel-free, lenses.

The association occupies 37 square kilometres (0.2 per cent of the land area) mostly in the Teith valley between Doune and Callander but also in the Earn valley around Crieff and in the middle of the Almond valley around Harrietfield.

Average annual rainfall varies from 1000 millimetres in the warm, moderately dry lowlands around Doune and Crieff to about 1500 millimetres in warm, wet lowlands around Callander and in the Almond valley. The dominant soils are brown forest soils with some humus-iron podzols in afforested areas and some gleys in the wet depressions between gravel mounds.

There is only one unit in the association, map unit 168. It consists mainly of freely drained brown forest soils developed on a gravel parent material which has been deposited as a series of mounds and terraces in valley bottoms. Slopes are variable, from gentle to steep, but most are strong. The topsoil has, usually, a sandy loam texture and is sometimes so thin and gravelly that stones are detrimental to arable agriculture and grass conservation where ploughing has been too deep. The steeper sides of mounds are also a limitation to arable agriculture as are some of the wet, intervening hollows. Much of the unit is used for limited arable cropping and permanent pasture (Lolio-Cynosuretum). The small areas of natural vegetation remaining are acid bent–fescue grassland (part of Achilleo-Festucetum tenuifoliae) and broadleaved woodlands. The latter comprise elmwood (Querco-Ulmetum glabrae), Atlantic oakwood and western highland birchwood (Blechno-Quercetum) and southern oakwood (Galio saxatilis-Quercetum). Some of the unit has been worked for sand and gravel extraction. In those areas restored to agriculture after such working, the topsoils are often of irregular thickness and occasionally are very thin with stones on the surface causing a bigger problem to grass conservation than is normal.

THE DREGHORN ASSOCIATION

(Map unit 169)

The soils of this association are developed on raised beach deposits derived mainly from Carboniferous igneous and sedimentary rocks with some Old Red Sandstone material. These deposits vary in texture from sandy loam or loamy sand to sandy gravel and frequently contain bands of fine sand or silt. In contrast to North Ayrshire where it was first mapped, the association is not extensive in Eastern Scotland and covers a total area of 48 square kilometres (0.2 per cent of the land area).

Only map unit 169 has been distinguished. It comprises freely drained brown forest soils and some poorly drained gleys with minor areas of imperfectly drained brown forest soils with gleying and small areas of noncalcareous regosols on windblown sand. The unit extends along the Fife coast from Fife Ness to Kirkcaldy, the most extensive area occurring between Anstruther and Leven. Smaller areas are also found in the Stirling, Bridge of Allan and Alloa districts. It occurs normally between the 15 metre- and 39 metre-levels on raised beach terraces and mounds with gentle slopes, although in one area it has been mapped as high as 61 metres above sea level. The average annual rainfall is 700 to 760 millimetres in the climate region of warm, moderately dry lowlands.

In the area around Fife Ness and southwards to Crail, gravel is generally conspicuous in the surface horizon which has a loamy sand texture. Further west, along the coast, the soils are of finer texture being sandy loams or fine sandy loams and gravel normally occurs only in the subsoil. The unit is used primarily for arable agriculture and there are few soil limitations. Although the coarse texture can sometimes lead to droughtiness, this is offset in most cases by the presence, at depth, of till which provides a valuable moisture reserve. Thus, some of the best arable land in Fife is located on the Dreghorn Association and much of map unit 169 produces a wide range of crops with high yields. Other areas produce only a moderate range and, where gravel is common, the yields are further restricted. Minor areas around Largo Bay, including areas of windblown sand, are used for rough grazing and are only moderately or marginally suited for grassland reclamation.

THE DULSIE ASSOCIATION

(Map units 172–175)

The Dulsie Association consists of soils developed on rudely stratified, gravelly loamy sands and sandy loams derived from acid schists of the Moinian but occasionally with a minor content derived from granite.

Though the parent material is similar to those of other associations developed on fluvioglacial materials, for example, the Corby or Auchenblae Associations, its degree of sorting and bedding is not as clearly defined. This poor grading is reflected by a consistently higher content of 20 to 25 per cent silt (2 to 50 μm) throughout the profile compared with the combined silt and clay fractions in fluvioglacial subsoils where 2 to 5 per cent is normal. Usually, the parent material of the Dulsie Association forms a veneer, 1 to 4 metres thick, overlying the local till or rock. Over much of the uplands south of Nairn, this drift is frequently sandwiched between a schist-derived till (the Arkaig Association) and; overlying mantle, 1 to 2 metres thick, of fluvioglacial sands and gravels of the Corby/Boyndie Associations; the sequence is usually covered by blanket peat. It is probable that the Dulsie parent material was formed during the penultimate stage of deglaciation and that the deposition was englacial.

Covering 189 square kilometres (0.8 per cent of the land area) the association within Eastern Scotland (Sheet 5) is restricted to the Moray Firth Lowlands and the Grampian Highlands. The altitude ranges from 30 metres to approximately 450 metres. Below about 250 metres, the landform is mainly non-rocky undulating lowlands and hills with gentle and strong slopes. In the Grampian Highlands the landform is more commonly a hummocky moraine though not necessarily restricted to the valleys. Average annual rainfall, related to altitude, varies from 800 millimetres in the climate region of fairly warm moderately dry lowlands and foothills to 1100 millimetres in the cool wet foothills and uplands.

The soils are dominated by freely drained humus-iron podzols and peaty podzols, frequently with an iron pan; occasionally they are imperfectly drained. A strongly indurated B horizon, with a well-developed platy structure, is characteristic in all the soils. Usually, this horizon is about 50 centimetres thick. The principal wet soils are peaty surface-water and ground-water gleys with minor areas of humic gleys associated with flushed sites. Small, rounded, bluish grey stones, generally less than 5 centimetres, form a large proportion of the stone content. Prominent very fine sandy loam caps on the stones, together with a well-developed fine pore system, are common in the C horizon. The colour of the parent material is typically a pale brown.

Arable and permanent pastures (Lolio-Cynosuretum) are widespread in the cultivated areas of the lowlands and fringing foothills. Soft rush pasture (the *Ranunculus repens–Juncus effusus* Community) and sedge mires occur on some of the gleys on the low ground. The greater part of the association, however, is dominated by dry and moist boreal heather moors (parts of Vaccinio-Ericetum cinereae) on the humus-iron podzols and peaty podzols with bog heather moors (Narthecio-Ericetum tetralicis) on the peaty gleys and peaty podzols of the higher slopes. Blanket bogs (Erico-Sphagnetum papillosi) accompanies the peat deposits. Many dry slopes on the foothills are afforested by Scots pine with Sitka and Norway spruce on the restricted wet soils.

Map unit 172   This is mainly a unit of humus-iron podzols. Minor components include some gleys and peaty podzols. The unit covers 79 square kilometres (40 per cent of the association) and is developed in a highly water-worked drift.

With a scattered distribution, areas are found usually at the junction of the undulating lowlands with the foothills of the adjacent 305-metre peneplain, for example, near Cawdor, the Divie and Dorback confluence and west of The Aird. Slopes are usually gentle and strong.

On the lower slopes, land use is generally long ley grassland with a limited arable rotation. With increasing altitude, climate limitations become more significant and arable rotations cease around 250 metres. Above this altitude, rough grazing, grouse-moor and forestry are the dominant land uses. Much of this ground below 350 metres is moderately suited to grassland reclamation.

Map unit 173   Consisting of peaty podzols with some peaty gleys and humus-iron podzols, this unit occurs on hill and valley sides with gentle and strong slopes. It is a minor unit which covers only 16 square kilometres (less than 10 per cent of the association) within the Spey and Dulnain valleys and near Ferness in the Findhorn valley.

Occurring mainly on convex or straight slopes, the unit is developed on a highly water-worked drift which is normally 2 or 3 metres thick and overlies either the local schist-derived till or bed-rock. An iron pan and induration are widespread.

Most of the area is planted to Scots pine and there is much natural regeneration on the surrounding grouse-moors in the southern areas where burning has been restricted in recent years. Some small areas are cultivated but are capable only of producing a narrow range of crops based on grass. Much of the lower uncultivated ground is suitable for grassland reclamation especially if the iron pan were disrupted by tine-ploughing. Some of the upper slopes are used as rough grazing; the grazings are of low value.

Map unit 174   This unit comprises peaty gleys and thin peat and occurs near Daviot on a hill side with gentle and steep slopes. It covers only 4 square kilometres (less than 5 per cent of the association) although other areas, too small to identify at the 1:250 000 scale, are distributed sporadically across the junction of the lowlands and the foothills.

The peaty gleys are developed on a highly water-worked drift which is often underlain by a schist-derived till at a depth of 2 metres or more. Situated on northern and north-western slopes, below a series of spring lines, the soils are subjected to surface-water gleying induced by the impermeable and strongly indurated B horizon. Additional problems are associated with the fluctuating ground-water table when the subsoils are disturbed; thixotropic conditions are created readily due to the high content of uniformly sized silt particles.

Some of the land is used as rough grazings though it is moderately or marginally suited to grassland reclamation. The organic surface horizons, induration and surface wetness are the main limitations to agricultural capability. Much of the area is planted to lodgepole pine and Sitka spruce.

Map unit 175   The unit consists of peaty podzols and occurs mainly as a hummocky moraine but also as a veneer of highly water-worked drift overlying the local till or bed-rock. Peaty podzols, gleyed above an iron pan, and imperfectly drained peaty podzols are included.

Covering 90 square kilometres (50 per cent of the association), it is situated mainly on the northern slopes of the plateau between the River Nairn and the River Dorback and on some slopes in mid-Strathspey (Plate 11).

The hummocks are elongated with each moraine field having a specific orientation. Occurring usually on the valley floors and the lower slopes, they are found also on gentle, convex upper slopes and summits.

On the lower flanks, there is substantial afforestation, mainly Scots pine and lodgepole pine, as well as a few limited areas of permanent pasture. Much of this lower ground is only marginally suited to grassland reclamation because of the thickness of the organic horizons, especially in the hollows. The remainder of the area is devoted to grouse-moors and rough grazing; the grazings are of low value. Climate, complex local topography and thickness of organic horizons are the primary limiting factors to agricultural capability.

Plate 11. *Strathspey, looking northwards from Ruthven Barracks near Kingussie. Undifferentiated alluvial soils of map unit 1 (Class 6.2) are bounded in the east by the discontinuous woodlands. Most are peaty, very poorly drained and subject to seasonal flooding. In the mid flood plain are the cultivated, mineral alluvial soils of the River Tromie delta (Class 3.2). The distant background is formed by the Western Cairngorm Mountains between Gleann Einich (upper left) and Glen Feshie (upper right). Subalpine soils of map unit 134 (Countesswells Association, Class 6.3) are characteristic of most of the southern slopes. The lower and gentler slopes of Glen Feshie comprise peaty podzols of map units 22 and 26 (Arkaig Association, Class 5.3). Above the flood plain (200 metres), the arable agriculture is associated largely with the humus-iron podzols of the terraces and outwash sands and gravels of map unit 100 (Corby Association, Class 4.2). The remaining low ground consists of peaty podzols, peat and peaty gleys of map unit 175 (Dulsie Association, Classes 5.2 and 5.3). Much of this area is moundy and is afforested. Isolated hills consist of peaty podzols of map unit 28 (Arkaig Association, Class 5.3).* Aerofilms.

THE DURNHILL ASSOCIATION

(Map units 181–183, 185, 187, 189, 192–195)

The soils of the Durnhill Association are developed on drifts derived from Dalradian quartzites and quartzose grits. Most of the drifts are shallow with abundant, mainly angular, quartzite stones, and textures of loamy sand or sandy loam. In the lowlands, however, on lower and concave slopes, the drift is usually a thicker stony till with textures ranging from loamy sand to sandy clay loam.

The drifts are often a characteristic light yellowish brown colour. Weathered quartzite underlies the drift locally and may contribute to the parent material of the soil.

Because of the resistant nature of the parent rocks, the association is restricted mainly to hills and mountains. In the north-east part of Eastern Scotland (Sheet 5), the hills include Mormond Hill, The Balloch, Millstone Hill and Ben Aigan. South-westwards from these areas, the association is found mainly on hills and mountains in Glenlivet and upper Donside and between Braemar and Pitlochry, for example, on Corryhabbie Hill, Beinn Iutharn Mhor and Beinn a'Ghlo. There is an outlying area farther to the south-west on Schiehallion. The association covers 297 square kilometres or 1.2 per cent of the land area. Altitudes range from near sea level to 1120 metres. Landforms include hills and valley sides, rugged hills and mountain summits with slopes ranging from gentle to very steep. Many of the steep slopes are scree-covered. Map units are non- to very rocky. The average annual rainfall is 700 to 1600 millimetres and the climatic regions range from warm, moderately dry lowlands on the coast to very cold, wet mountains. The plant communities include boreal heather moors (Vaccinio-Ericetum cinereae), alpine azalea–lichen heath (Alectorio-Callunetum vulgaris) and upland and mountain blanket bogs (part of Erico-Sphagnetum papillosi and Rhytidiadelpho-Sphagnetum fusci).

Peaty podzols, alpine soils and peaty rankers are the most extensive soils. The peaty podzol is one of the most striking soils in Eastern Scotland (Sheet 5), the grey 'sugary' quartz-rich E horizon contrasting strongly with the brightly coloured, reddish yellow Bs horizon. Indurated B horizons occur on the less stony parent materials. The organic-matter content of the black humose Bh horizons of alpine soils tends to be lower than on soils developed on less acid parent materials. Peaty rankers are common on stabilized scree, the black, peaty horizons being underlain by interlocking quartzite boulders with black humose interstitial material. The general absence of brown forest soils is a consequence of the low nutrient content and acid nature of the soil parent materials.

Map unit 181   This consists of noncalcareous and peaty gleys with some humic gleys and peat. It covers 8 square kilometres (3 per cent of the association) near the Hill of Dudwick and in the Keith–Cullen districts. The unit is non-rocky and occurs on undulating lowlands and hills with concave, gentle and strong slopes.

The soils are developed on a stony till which ranges in texture from loamy sand to sandy clay loam. They are weakly structured and the topsoils of the noncalcareous gleys are dark-coloured.

Plant communities include permanent pastures (Lolio-Cynosuretum) and rush pastures on noncalcareous gleys, bog heather moors (Narthecio-Ericetum tetralicis) and sedge mires on peaty and humic gleys, and blanket bogs (Erico-Sphagnetum papillosi) on peat. In the Hill of Dudwick area, the land is cultivated and produces a moderate range of crops with average yields. Elsewhere, the

Plate 12. *The head of Glen Shee, looking north to the Cairngorm Mountains. In the foreground, the soils of the valley are peat and peaty gleys of map unit 523 (Tarves Association); the grazings are of moderate quality (Class 6.2). White quartzite hills in the middle distance stand out in sharp contrast; the soils are rankers and lithosols of map unit 195 (Durnhill Association). There, the land is of very limited agricultural value (Class 7) and is used for deer-stalking, for grouse- and ptarmigan-shooting and, as the tracks and ski-runs show, for recreation.* Cambridge University Collection: copyright reserved.

land use ranges from arable agriculture to uncultivated areas which are moderately suited for grassland reclamation.

Map unit 182   This unit comprises peaty podzols with some freely and imperfectly drained humus-iron podzols, gleys and subalpine soils. It is the most extensive in the association and covers 134 square kilometres (45 per cent of the association). The unit occurs on hills, valley sides and undulating lowlands with mainly gentle and strong, non-rocky slopes. There are steep slopes locally, some of which are scree-covered.

Most soils are formed on a stony or very stony, loamy sand or sandy loam drift, but on Windy Hill near Fyvie the parent material is a gravel. The latter may be of Pliocene age and consists mainly of quartzite and quartz-schist pebbles and gravel. Most of the humus-iron podzols and some of the peaty podzols in the lowlands have been cultivated. These soils display the very dark, often humose, surface horizons characteristic of cultivated podzols.

The most extensive plant communities are dry and moist boreal heather moors (parts of Vaccinio-Ericetum cinereae). There are arable and permanent pastures (Lolio-Cynosuretum) and rush pastures in cultivated areas. Most of the cultivated

100

land produces a moderate or narrow range of crops, the latter based primarily on long ley grassland. On uncultivated ground, used mainly for rough grazing, forestry and grouse-shooting, some areas are moderately suitable for grassland reclamation. The rough grazing has a low grazing value.

Map unit 183   This unit consists of peaty podzols and peat, with some subalpine and alpine soils, peaty gleys and rankers. It occupies 4 square kilometres (1 per cent of the association) in the Glenlivet area on landforms similar to those of *map unit 182* but is restricted to altitudes above 350 metres.

The vegetation is mainly moist boreal heather moor (part of Vaccinio-Ericetum cinereae) and bog heather moors (Narthecio-Ericetum tetralicis) on peaty podzols, and lowland, upland and mountain blanket bogs (parts of Erico-Sphagnetum papillosi and Rhytidiadelpho-Sphagnetum fusci) on peat. The land is used for rough grazing, forestry and grouse-shooting. Altitude and exposure limit agriculture to rough grazing with low grazing values, sometimes of a seasonal nature.

Map unit 185   Consisting of peaty podzols and peat with some subalpine soils and peaty gleys, the unit occupies 1 square kilometre (less than 1 per cent of the association) in two areas south-west of Braemar. The landform is specific, non-bouldery, moundy valley moraines, pointed or round-topped with gentle to steep slopes and associated basins of peat. Altitude ranges from 450 to 650 metres.

The soils on the moraines are mainly peaty podzols, with subalpine soils at the highest elevations and on exposed tops of mounds; the vegetation is dry and moist boreal heather moors (parts of Vaccinio-Ericetum cinereae). Basins between the moraines have upland and mountain blanket bogs on peat (part of Erico-Sphagnetum papillosi and Rhytidiadelpho-Sphagnetum fusci) and bog heather moors (Narthecio-Ericetum tetralicis) on the local peaty gleys. The land is used for rough grazing which has a low grazing value.

Map unit 187   This map unit comprises peaty podzols and peaty rankers with occasional peat and subalpine soils. It occupies 2 square kilometres (less than 1 per cent of the association) in Glenlivet and upper Donside and occurs on hills with strong to very steep slopes. The unit is slightly rocky, with scree and occasional rock outcrops.

Dry and moist boreal heather moors (parts of Vaccinio-Ericetum cinereae) form the dominant vegetation. The land is used for rough grazing, forestry and grouse-shooting. Rock, slope and climate restrict agriculture to the rough grazings which are of low grazing value.

Map unit 189   Comprising peaty rankers with some peaty podzols, peat and subalpine soils, the unit occupies 4 square kilometres (1 per cent of the association) in upper Donside and south of Braemar in Glen Clunie. The landforms are similar to those of *map unit 187* but there is more scree and rock outcrop. As with *map unit 187*, scree occupies a larger area than rock outcrops and usually predominates on steep slopes.

The vegetation and land use are similar to those of *map unit 187*, but map unit 189 is too rocky for forestry, and agriculture is further restricted to mainly seasonal, very poor, rough grazings.

Map unit 192   This map unit consists of subalpine soils with some peat and occasional peaty podzols, gleys, rankers and alpine soils. It occupies 26 square

101

kilometres (9 per cent of the association), mainly on and near Beinn a'Ghlo with similar areas near Braemar and on Schiehallion. The unit occurs on broad, rounded summits, ridge tops and hill sides with gentle and strong and locally steep slopes. It is usually non- or slightly rocky, but there are a few moderately to very rocky areas and scree-covered slopes.

Peat ranges from a major component co-dominant with subalpine soils to a minor component on rocky and steep slopes.

Plant communities include lichen-rich boreal heather moor (part of Vaccinio-Ericetum cinereae) and alpine azalea–lichen heath (Alectorio-Callunetum vulgaris) on subalpine soils, rankers and alpine soils, and upland and mountain blanket bogs (part of Erico-Sphagnetum papillosi and Rhytidiadelpho-Sphagnetum fusci) on peat. The land is used for rough grazings which have low grazing values; some of the grazings are of a seasonal nature. Altitude and climate, especially exposure, are the main limitations.

Map unit 193   Peat and alpine soils with some subalpine soils and rankers form this unit. It is closely related to *map unit 192*, and occurs on similar landforms but with alpine soils replacing the subalpine soils. The unit occupies 40 square kilometres (13 per cent of the association), mainly in upper Donside and Glenlivet and on the mountains south-west of Braemar.

Plant communities include alpine azalea–lichen heath (Alectorio-Callunetum vulgaris) on alpine soils and mountain blanket bog (Rhytidiadelpho-Sphagnetum fusci) on peat. The land is used for rough grazing though the very severe limitations of exposure and soil restrict the grazing to a mainly very poor and seasonal nature.

Map unit 194   This unit consists of alpine soils with some peat, rankers, lithosols and subalpine soils. It occupies 10 square kilometres (3 per cent of the association), mainly on and near Corryhabbie Hill south of Dufftown. The unit occurs on rounded mountain summits and mainly convex, upper slopes, where it ranges from non- to moderately rocky with a few very rocky areas of scree and rock outcrop. Slopes range from gentle to steep.

The vegetation on mineral soils is mainly alpine azalea–lichen heath (Alectorio-Callunetum vulgaris) with mountain blanket bog (Rhytidiadelpho-Sphagnetum fusci) on peat. Land use and the capability for agriculture are the same as for *map unit 193*.

Map unit 195   Similar to *map unit 194* but rockier, this unit is usually dominated by rankers and lithosols, with some alpine and subalpine soils and peat. It occupies 68 square kilometres (23 per cent of the association), mainly on mountains between Braemar and Pitlochry (Plate 12) and on Schiehallion. The unit occurs on mountain summits and sides with gentle to very steep slopes.

In less rocky areas, alpine soils are sometimes dominant. Many of the steep slopes are extensively scree-covered; boulder-fields are common and there are some precipitous rock faces in corries.

The plant communities include alpine azalea–lichen heath (Alectorio-Callunetum vulgaris), stiff sedge–fescue grasslands (the *Carex bigelowii–Festuca vivipara* Association) and blaeberry heath (the *Rhytidiadelphus loreus–Vaccinium myrtillus* Community). Land use and the capability for agriculture are the same as for *map unit 193*.

THE ECKFORD/INNERWICK ASSOCIATIONS

(Map units 199 and 200)

The soils in this group are developed on fluvioglacial sands and gravels derived mainly from Upper Old Red Sandstone sediments. In Eastern Scotland (Sheet 5), where only soils of the Eckford Association have been mapped, the parent material also includes some andesitic lava from the Ochil Hills. Covering 97 square kilometres (0.4 per cent of the land area), these soils occur in the moderately dry lowlands with an average annual rainfall of 700 to 760 millimetres. Two map units have been distinguished.

Map unit 199 Developed on sand, the unit comprises noncalcareous gleys with small areas of mineral alluvial soils and humus-iron podzols. It is a minor unit, covering only 3 square kilometres (less than 5 per cent of the association) around the shores of Loch Leven at an altitude of 110 metres.

Land use is restricted to a moderate range of crops with average yields. Wetness is the main limitation.

Map unit 200 The more extensive map unit 200, covering 94 square kilometres (95 per cent of the association), comprises freely and imperfectly drained humus-iron podzols with some gleys and mineral alluvial soils. It occurs mainly in the Howe of Fife and on the plain of Kinross over an altitude range of 40 to 135 metres. Although they are not distinguished at the 1:250 000 scale, the unit contains areas of two soil series, one developed mainly on sand and occurring on gentle slopes and the other developed mainly on gravel and found on low mounds. Both give rise to podzols, most of which are under cultivation or support coniferous woodland as in the area immediately north of Ladybank. Much of the area is capable of producing a moderate range of crops, especially where the top-soils are thicker and where the soils are imperfectly drained. Coarse texture, resulting in low moisture- and nutrient-retention capacity, causes moderate limitations to arable cultivation in many soils. In one area, additional restrictions imposed by gravel limit land use to a narrow range of crops based primarily on long ley grassland.

THE ELGIN ASSOCIATION

(Map units 201–203)

Soils of the Elgin Association are developed on drifts derived from Upper Old Red Sandstone sediments; some drifts derived from sandstones of Middle Old Red Sandstone or Permo-Triassic age have been included. The sediments are mainly grey, yellow and red sandstones, often coarse and pebbly. The drift, mainly a sandy loam till, is up to 6 metres thick; the characteristic pale reddish brown colour usually darkens and reddens with depth and the clay content increases. An indurated horizon occurs commonly at approximately 50 centimetres below the surface. Sandstones are the most common stones in the drift but the proportion of acid metamorphic rocks rises as the Moinian rocks are approached in the south. Freely drained humus-iron podzols are the dominant soils; other soils include peaty podzols and noncalcareous, peaty and humic gleys.

The association covers 62 square kilometres (0.3 per cent of the land area) and is restricted to the Moray Firth Lowlands between Buckie and Kinloss. In general, it

103

occurs on the higher ground of the plain, protruding through the mantle of fluvioglacial sands and gravels. The largest areas are on the ridges of Monaughty and Quarrywood west of Elgin and on the coastal ridge between Burghead and Lossiemouth. These areas lie below 150 metres. Another large area occurs to the south-west of Elgin at Teindland where the ground rises to 340 metres.

Landforms include undulating lowlands with gentle and strong slopes, and hills and valley sides with gentle and steep slopes. Rock outcrops are rare. Average annual rainfall ranges from 630 to 900 millimetres and the climatic regions vary from warm moderately dry lowland to fairly warm moderately dry lowland and foothill.

Map unit 201   This unit consists of noncalcareous and peaty gleys with some humic gleys. It occurs in depressions and receiving sites with gentle slopes in lowland areas. Covering 4 square kilometres (5 per cent of the association), the unit exists only at two small areas. The larger area, at Teindland to the south-east of Elgin, contains both noncalcareous and peaty gleys; the other area lies south of Alves and consists entirely of poorly drained noncalcareous gleys.

Plant communities include arable and permanent pastures (Lolio-Cyno-suretum), rush pastures and sedge mires with bog heather moors (Narthecium-Ericetum tetralicis) on the peaty gleys.

Some of the land is cultivated and capable of producing a narrow range of crops and some is in permanent pasture and rough grazings. The main limiting factor is wetness although the uncultivated land is marginally improvable for grassland.

Map unit 202   This unit, covering 57 square kilometres (90 per cent of the association), is the largest in the association and occurs throughout the association area. Humus-iron podzols are the dominant soils; they are largely free-draining and cultivated. At higher elevations, imperfectly drained podzols become common and afforestation is widespread, for example, on the north-facing slope of Monaughty. Peaty podzols and gleys are minor component soils.

It occurs in undulating lowlands and hills where the slopes are predominantly gentle and strong although locally, as in Glen Rothes, steep slopes are present. The landscape is mainly non-rocky but there are rock outcrops along the ridge between Alves and Elgin.

Plant communities include arable and permanent pastures (Lolio-Cyno-suretum) and eastern highland oak and birchwood (Trientali-Betuletum pendulae) on the podzols and locally, rush pastures and sedge mires on the gleys.

Arable agriculture and forestry, mainly coniferous plantations, are the main land uses. Climate is the principal limiting factor affecting the agriculture and it becomes progressively more restrictive with altitude. On the higher ground, agriculture is limited to grassland with uncultivated areas only moderately suited to reclamation.

Map unit 203   The only area of this unit occurs on the higher ground at Teindland, north of Rothes, where it occupies 1 square kilometre (less than 5 per cent of the association). It lies between 200 and 300 metres where the hills are mainly gently sloping and non-rocky.

Peaty podzols, in places transitional to humus-iron podzols, are the dominant soils. Peat and gleys are associated with local depressions.

The main plant communities are boreal and Atlantic heather moors (Vaccinio-Ericetum cinereae and Carici binervis-Ericetum cinereae respectively) on the podzols with blanket bogs (Erico-Sphagnetum papillosi) and rush pastures and sedge mires on the peat and gleys.

THE ETHIE ASSOCIATION

(Map unit 204)

Soils of the Ethie Association are developed on drifts derived from sandstones of the Middle Old Red Sandstone and acid gneisses of the Moine, and possibly of the Lewisian. They occupy 16 square kilometres in the Black Isle and Easter Ross (less than 0.1 per cent of the land area). The distribution corresponds to the coastal band of Moine gneiss which extends discontinuously from Rosemarkie to the Hill of Nigg and includes the Sutors of Cromarty. There is only one map unit and the description is combined with that of the association.

The landform is one of low hills, often in the form of long ridges. Slopes are normally strong but locally steep or very steep, especially seaward slopes. The hills north of Rosemarkie rise to 203 metres at Hill of Nigg and 220 metres at Callachy Hill. There are sporadic rock outcrops.

Climatically, the association lies in the region of warm and fairly warm, moderately dry lowlands. Average annual rainfall is between 630 and 750 millimetres.

The parent material is mainly a brownish, stony drift, often less than 50 centimetres thick and usually with a sandy loam texture; the texture ranges occasionally from loamy sand to loam. It is derived largely from sandstone but incorporates a variable proportion of acid gneisses. The amount of gneiss normally increases with depth and, in the very shallow soils, the rubbly bed-rock is the main component of the drift. Colluvial material often infills local depressions and may overlie till of the Cromarty Association.

Freely drained humus-iron podzols are the dominant soils of the unit and occupy the hill tops and convex slopes. The podzols developed on shallow drift normally have dark reddish brown B horizons whereas the B horizons in the deeper drift are generally yellowish brown. Induration occurs sporadically. Iron pans are not normally associated with these soils. Imperfectly drained podzols, on the hill slopes, become progressively wetter downslope. Minor patches of poorly drained gleys occur in the valley floors. Although mainly noncalcareous, a few are calcareous below the topsoil which is usually 40 centimetres thick.

Most of the unit is utilized for mixed farming or is afforested. Although steep slopes and the shallow, stony soils restrict much of the agricultural capability to grass production, minor areas are suited to a moderate or narrow range of crops.

The plant communities include arable and permanent pastures (Lolio-Cynosuretum) and dry Atlantic heather moor (part of Carici binervis-Ericetum cinereae) on the humus-iron podzols with soft rush pastures (the *Ranunculus repens-Juncus effusus* Community) and sedge mires on the gleys.

THE FORFAR ASSOCIATION

(Map units 237–239)

The Forfar Association, covering a total area of 375 square kilometres (1.5 per cent of the land area) is developed mainly on water-sorted drifts and on colluvial material derived from Lower Old Red Sandstone sediments. The drifts are believed to have been reworked and re-sorted in post-glacial times. The association is found throughout, or is contiguous with, areas of the Balrownie Association to which it is closely related. It occurs widely in lowland Strathmore, extending south-westwards to the outskirts of Crieff. The altitudinal range is from 15 to 165 metres with the greater part lying between 45 and 120 metres.

Like Balrownie, the Forfar Association occurs in regions ranging from fairly warm, moderately dry lowlands with an average annual rainfall of 1500 millimetres, to fairly warm, wet lowlands and foothills with 2000 millimetres rainfall. Three units have been mapped.

Map unit 237   Covering an area of 66 square kilometres (18 per cent of the association), the unit comprises brown forest soils with some humus-iron podzols and noncalcareous gleys. It extends from the south-west end of Strathmore to the vicinity of Crieff, occurring on undulating lowlands with gentle and strong slopes. The dominant soil has a sandy loam or loam topsoil on subsoil horizons of sandy loam or loam texture which are underlain at a depth of 60 centimetres or more by a reddish brown loam or sandy clay loam drift.

Most of the unit is devoted to arable agriculture capable of a wide or moderate range of crops. Rapid natural drainage associated with the coarse, upper textures allows access to the land at most seasons and the underlying drift ensures nutrient storage and reduces susceptibility to drought.

Map unit 238   The unit covers only 7 square kilometres (2 per cent of the association) and occurs on undulating lowlands with gentle slopes at the south-west end of Strathmore. It comprises noncalcareous gleys with some peaty and humic gleys.

Although most of the unit is cultivated, it is restricted by wetness and slope limitations to producing a moderate range of crops with average yields A few small, uncultivated areas are highly suited to grassland reclamation.

Map unit 239   This is the most extensive of the three map units and covers a total area of 302 square kilometres (80 per cent of the association), much of it occurring between the Howe of the Mearns and Perth. It is also found in several river valleys and drainage channels and in the broad colluvial hollows lying to the north of Forfar. It is closely related to *map unit 237* and occurs on similar topography, but the dominant components are humus-iron podzols, with brown forest soils and noncalcareous gleys as minor constituents. Because of variations in the nature of the parent material and in the degree of water-sorting, a wide range of profiles is found within the unit. The dominant soil, however, has a textural profile almost similar to that of *map unit 237* with a sandy loam or loam topsoil on a sandy loam or loamy sand subsoil underlain by a loam or sandy clay loam.

As with *map unit 237*, this unit is devoted to highly productive arable agriculture with a wide, or moderate, range of crops.

THE FOUDLAND ASSOCIATION
(Map units 240, 241, 243–246, 248, 250, 252, 253, 255–258)

The soils of the Foudland Association are developed on drifts derived from slates, phyllites and other weakly metamorphosed argillaceous rocks, including andalusite-schists, fine-grained mica-schists, and black and knotted schists of the Dalradian series. The drift is a yellowish brown to olive-brown fine sandy loam or loamy fine sand, with many platy stones mainly less than 20 centimetres across.

The soil has a smooth feel due to the considerable amounts of silt (2–50 µm). Locally, weathered rock underlies the drift and thus has contributed to the parent material.

The association is extensive in the north-east area of Eastern Scotland (Sheet 5), mainly in the country around Huntly and Turriff. Other areas in the north-east are the Correen and Ladder Hills, the Keith and Dufftown districts, Glenlivet and Strathavon, the Cabrach and upper Donside, and the Glen Clunie–Glas Maol–Glen Shee districts. Farther south, the association occurs in the Ben Lawers area; it also stretches from Dunkeld to the Sma' Glen, and from Crieff through Aberfoyle to Loch Lomondside. The association covers 1310 square kilometres or 5.3 per cent of the land area. Elevations range from sea level to the summit of Ben Lawers at 1215 metres. The landforms include undulating lowlands, valley sides, rounded hills, hummocky valley moraines and mountains. Slopes range from gentle to very steep. Although the map units range from non-rocky to very rocky, most of the area is non-rocky. This is partly a consequence of the nature of the rock which weathers fairly readily.

The average annual rainfall is 800 to 1100 millimetres in the main area, that is around Huntly and Turriff, but reaches 2400 millimetres on Ben Lawers. The climatic region of most of the main area comprises fairly warm, moderately dry lowlands and foothill, but elsewhere it ranges to very cold wet mountains. The main plant communities are arable and permanent pastures (Lolio-Cynosuretum) and boreal heather moors (Vaccinio-Ericetum cinereae). Others include acid bent–fescue grassland (part of Achilleo-Festucetum tenuifoliae), common white bent grassland (part of Junco squarrosi-Festucetum tenuifoliae), sharp-flowered rush pastures (Potentillo-Juncetum acutiflorae), upland and mountain blanket bogs (part of Erico-Sphagnetum papillosi and Rhytidiadelpho-Sphagnetum fusci) and alpine azalea–lichen heath (Alectorio-Callunetum vulgaris).

The parent materials of the association are acid and low in bases, so that in the prevailing climate brown forest soils are local and restricted to the lowlands and to steep slopes where the soils are subject to rejuvenation. The most extensive soils are humus-iron podzols. In the lowlands and foothills where many soils are cultivated, most fields have wire fences rather than stone walls. This results from the scarcity of suitable dyke stones unearthed during cultivation. Another characteristic of the association is that gleys are relatively uncommon, probably because the soil and the bed-rock, which is often shattered and steeply folded, allow ready percolation of drainage water. Induration is fairly common in humus-iron and peaty podzols but is usually weak or very weak. Peaty podzols and uncultivated humus-iron podzols developed on drifts derived from black schists have characteristic bluish grey E horizons. Such soils are commonest in the Strathdon, Glenlivet and Dufftown areas. The argillaceous sediments forming the parent materials of the association have adequate amounts of cobalt and copper, although applications of lime for improved grazings of hill land can induce deficiencies of these trace elements.

Map unit 240 This unit consists of brown forest soils with some brown rankers and noncalcareous gleys. It occupies 18 square kilometres (1 per cent of the association) in the Trossachs and Glen Artney with a small area in Strathavon near Tomintoul. Most of the unit is slightly rocky and occurs on hill summits and on valley sides with gentle to steep slopes. The soils are shallow.

The vegetation includes acid bent–fescue grassland (part of Achilleo-Festucetum tenuifoliae), herb-rich boreal heather moor (part of Vaccinio-Ericetum cinereae) and boreal juniper scrub (Trientali-Juniperetum communis).

107

The land is used for rough grazings and forestry. Some areas are moderately suited to grassland reclamation, the remainder having high grazing value. Climate, slope and soil are the principal limitations.

Map unit 241   Noncalcareous and peaty gleys, with some humic gleys, peat and humus-iron and peaty podzols are the soils of this unit. It covers 103 square kilometres (8 per cent of the association) and is widespread but local in the lowlands and foothills throughout the area of the association. The landforms are hollows, basins and concave, gently and strongly sloping lower slopes. The unit is non-rocky. The soils are formed on grey fine sandy loam or loam till which usually has 25 to 50 per cent silt (2–50 μm).

The plant communities include arable and permanent pastures (Lolio-Cynosuretum), rush pastures, sedge mires, and flying bent grassland (part of Junco squarrosi-Festucetum tenuifoliae). Most of the land is cultivated and produces a moderate range of crops with average yields. Wetness is the principal limitation.

Map unit 243   The soils of map unit 243 are humus-iron podzols with some brown forest soils, noncalcareous and peaty gleys, peaty podzols and peat. By far the most extensive unit of the association, it covers 878 square kilometres (67 per cent of the association) throughout the area of the association, especially in the Huntly–Turriff district. The map unit occurs on undulating lowlands, rounded hills and valley sides. Slopes tend to be smooth and long. They are usually gentle and strong, locally steep. The unit is non-rocky.

Most of the humus-iron podzols are cultivated. They are usually freely drained and have a friable Bs horizon which is usually less stony than the underlying, weakly indurated B horizon. The B horizons tend to be yellower than those of the allied associations with hues ranging from 10YR to 5Y. This is most marked in soils on black schists, where they are olive-brown or even olive. Gleys and peat are found in receiving sites, mainly hollows and basins.

The vegetation is mainly arable and permanent pastures (Lolio-Cynosuretum) with some dry boreal heather moor (part of Vaccinio-Ericetum cinereae) on uncultivated humus-iron podzols, and acid bent-fescue grassland (part of Achilleo-Festucetum tenuifoliae). The land is mostly cultivated and produces a moderate range of crops, though there are minor areas limited to a narrow range based on long ley grassland. Uncultivated podzols are moderately suited to grassland reclamation.

Map unit 244   This unit comprises peaty podzols with some humus-iron podzols, noncalareous and peaty gleys, subalpine soils and peat. It covers 73 square kilometres (6 per cent of the association) in the foothills and uplands, the areas including upper Donside and the Correen Hills. The unit occurs on rounded hills with gentle to steep slopes and is non-rocky. Some peaty podzols, especially on convex slopes, have freely drained E horizons. Iron pans are not normally found or are only weakly formed.

The vegetation includes boreal heather moors (Vaccinio-Ericetum cinereae) and common white bent grassland (part of Junco squarrosi-Festucetum tenui-foliae) on podzols and subalpine soils and bog heather moors (Narthecio-Ericetum tetralicis) on peaty gleys. The land is used for rough grazing, grouse-shooting and forestry. Grazing values are low. Some areas are moderately suited to grassland reclamation. Climate and slope are the dominant limitations.

Map unit 245 Consisting of peaty podzols and peat, with some humus-iron podzols, subalpine soils and peaty and noncalcareous gleys, this unit occurs usually at higher elevations and on rather gentler slopes than *map unit 244*. It covers 121 square kilometres (9 per cent of the association), mainly in the Cabrach, Ladder Hills and upper Donside areas. The unit occurs on non-rocky rounded hills with gentle, strong and locally steep slopes and on plateaux. Peat tends to be most common on plateaux and on upper and gentler slopes whereas the podzols conform to an altitudinal zonation with the humus-iron podzols confined to the lowest elevations.

The plant communities include boreal heather moors (Vaccinio-Ericetum cinereae) on peaty podzols, and blanket bogs (Erico-Sphagnetum papillosi) and mountain blanket bog (Rhytidiadelpho-Sphagnetum fusci) on peat. The land is used for rough grazing, grouse-moor and forestry. Grazing values are low.

Map unit 246 This map unit consists of peaty gleys with some peat, peaty rankers and humic gleys. It covers 9 square kilometres (less than 1 per cent of the association) in the Trossachs. It occurs on gentle and strong, lower hill slopes. The relief is rock-controlled but with few rock outcrops.

The soils are formed on shallow drifts, some of which are colluvial. The vegetation includes moist Atlantic heather moor (part of Carici binervis-Ericetum cinereae), bog heather moors (Narthecio-Ericetum tetralicis) flying bent bog (part of Erico-Sphagnetum papillosi) and sedge mires. The land is used for rough grazing which has a low grazing value.

Map unit 248 This map unit consists of humus-iron and peaty podzols with some peaty gleys and peat. It covers 5 square kilometres (less than 1 per cent of the association) in the Glen Artney area. It occurs on valley floors and lower valley sides and has a characteristic landform of hummocky moraine with associated channels and hollows.

The soils are formed on moraine derived from slates, phyllites and argillaceous schists. The podzols occur on the mounds and the peaty gleys and peat in the channels and hollows.

The plant communities include acid bent–fescue grassland (part of Achilleo-Festucetum tenuifoliae) and permanent pastures (Lolio-Cynosuretum) with some Atlantic heather moor (Carici binervis-Ericetum cinereae) on the podzols, and rush pastures and sedge mires on the peaty gleys and peat. The land is used for rough grazing with some forestry, but is moderately suited to grassland re-clamation.

Map unit 250 The soils of this map unit are brown forest soils and brown rankers with some noncalcareous and peaty gleys and humus-iron podzols. The unit covers 7 square kilometres (less than 1 per cent of the association) in the Callander and Aberfoyle areas. It occurs on hill and steep valley sides and is moderately rocky.

The soils are formed on shallow, stony drifts with humus-iron podzols restricted to higher elevations, and the gleys restricted to flushes on the slopes.

Plant communities include acid bent–fescue grassland (part of Achilleo-Festucetum tenuifoliae) and hazelwood (part of Querco-Ulmetum glabrae) and eastern highland oak and birchwood (Trientali-Betuletum pendulae) on the brown forest soils and brown rankers, and rush pastures and sedge mires on the gleys. The land is used for rough grazing and forestry.

Map unit 252 This map unit consists of humus-iron and peaty podzols and peaty rankers with some noncalcareous and peaty gleys and peat. It covers 17 square kilometres (1 per cent of the association) in the Aberfoyle, Callander and Sma' Glen areas. Though landform and parent materials are similar to *map unit 250*, this unit occurs at higher elevations with podzols and peaty rankers replacing the brown forest soils and brown rankers.

The vegetation includes boreal heather moors (Vaccinio-Ericetum cinereae), some Atlantic heather moor (Carici binervis-Ericetum cinereae), and common white bent grassland (part of Junco squarrosi-Festucetum tenuifoliae) on the podzols and peaty rankers, and blanket bogs (Erico-Sphagnetum papillosi) on peat. The land is used for rough grazing. Grazing values are moderate or low.

Map unit 253 This map unit comprises peaty gleys and peaty rankers with some peat, humic gleys and peaty and humus-iron podzols. It covers 19 square kilometres (2 per cent of the association) in the Trossachs, Callander and Killin areas. Occurring on gently ridged hills with gentle to steep slopes, it is slightly rocky.

The soils are formed on shallow, mainly colluvial, drifts or on patches of lodgement till on lower concave slopes. The unit is similar to *map unit 252*, but wetter.

Plant communities include northern and moist Atlantic heather moors (parts of Carici binervis-Ericetum cinereae), bog heather moors (Narthecio-Ericetum tetralicis), heath grass–white bent grassland (part of Junco squarrosi-Festucetum tenuifoliae) and soft rush pasture (the *Ranunculus repens–Juncus effusus* Community). The land is used for rough grazing and forestry. Grazing values are mainly low.

Map unit 255 Consisting of subalpine soils with some rankers, peat and alpine soils it covers 19 square kilometres (2 per cent of the association), mainly in the Ben Lawers and Meall nan Tarmachan ranges. The unit occurs on mountain summits with gentle to very steep slopes and ranges from non- to very rocky.

Most soils are formed on shallow, stony drifts. The degree of rockiness controls the frequency of rankers, which become dominant in very rocky areas. Most subalpine and alpine soils are freely drained.

The vegetation includes stiff sedge–fescue grasslands (the *Carex bigelowii–Festuca vivipara* Association), mountain blanket bog (Rhytidiadelpho-Sphagnetum fusci), lichen-rich boreal heather moor (part of Vaccinio-Ericetum cinereae) and alpine azalea–lichen heath (Alectorio-Callunetum vulgaris). The land is used for rough grazing which is occasionally of a very poor and seasonal nature. The grazing values are low.

Map unit 256 This map unit consists of peat and alpine soils with some subalpine soils. It occupies 24 square kilometres (2 per cent of the association) mainly in upper Donside, the Ladder Hills and in the Glen Clunie–Glas Maol area. The unit occurs above 600 metres on rounded hill summits and upper gentle and strong slopes. It is non-rocky, occasionally slightly rocky.

The alpine soils are mainly on summits and convex slopes, with subalpine soils confined to lower ground. These soils are freely drained with organic-rich subsoils which are black in alpine soils and dark reddish brown in subalpine soils. Peat is usually dominant on plateaux and cols.

The plant communities include mountain blanket bog (Rhytidiadelpho-Sphagnetum fusci) on peat, with alpine azalea–lichen heath (Alectorio-

Callunetum vulgaris) on alpine soils, and lichen-rich boreal heather moor (part of Vaccinio-Ericetum cinereae) on subalpine soils. The land is used for rough grazing and grouse-shooting. The grazing is mainly very poor and of a seasonal nature.

Map unit 257 This unit has alpine soils with some peat, rankers and subalpine soils. It occupies 13 square kilometres (1 per cent of the association) in the Ben Lawers range and in the Glen Clunie–Glas Maol–Glen Shee area. It occurs on mountain summits with gentle and strong slopes and is non- or slightly rocky, except for the corrie walls.

Most of the alpine soils are freely drained. They usually lack or have only weakly expressed organic-rich subsoils, but have a brown Bs horizon. Peat and alpine gleys occur locally in hollows.

The plant communities include alpine azalea–lichen heath (Alectorio-Callunetum vulgaris) and stiff sedge–fescue grassland (the *Carex bigelowii–Festuca vivipara* Association), with mountain blanket bog (Rhytidiadelpho-Sphagnetum fusci) on peat. Species-rich communities occur in the Glas Maol area, associated with outcrops of limestone. Land use is limited mainly to very poor, seasonal, rough grazing.

Map unit 258 This map unit consists of rankers and lithosols with some alpine and subalpine soils. It covers 4 square kilometres (less than 1 per cent of the association) in the Ben Lawers range. It is the rocky equivalent of *map unit 257* and occurs on rocky mountain summits and ridges including corrie walls and rock cliffs.

The soils are formed on very shallow stony drifts. The vegetation includes fescue–woolly fringe-moss heath (Festuco-Racomitrietum lanuginosi), and stiff sedge–fescue and upland bent–fescue grasslands (respectively the *Carex bigelowii–Festuca vivipara* Association and part of Achilleo-Festucetum tenuifoliae). Species-rich communities also occur, associated with narrow outcrops of lime-rich rock. Land use is restricted to very poor, seasonal rough grazing.

THE FRASERBURGH ASSOCIATION

(Map unit 259)

Soils of the Fraserburgh Association, previously described in Aberdeenshire, East Lothian and Caithness, are developed on windblown shelly sands underlain by raised beach deposits. In Eastern Scotland (Sheet 5), the association has been mapped in only two districts, in north-east Aberdeenshire and in Fife, a total area of 6 square kilometres (less than 0.1 per cent of the land area). In the north-east, between Fraserburgh and Rosehearty and immediately south of Rattray Head, it occurs on level to gently sloping terraces at altitudes from 15 to 24 metres. In Fife, the association is limited to an area of 4 square kilometres on the shores of Largo Bay near Elie where it occurs at an altitude of 15 metres on raised beach terraces, and on dunes with gentle and steep slopes. Only one map unit has been distinguished.

Map unit 259 consists mainly of brown calcareous soils with some calcareous regosols. The dominant soil has a dark brown shelly loamy sand about 20 centimetres thick on a dark yellowish brown loamy sand about 25 centimetres thick on shelly coarse sand.

111

In the north, the unit occurs in a region of warm moderately dry lowlands with an average annual rainfall range of 700 to 760 millimetres; in the south, it is found on warm, moderately dry lowlands with 650 millimetres rainfall.

The plant communities are mainly northern dunes (Elymo-Agropyretum boreo-atlanticum and Elymo-Ammophiletum) and dune pastures (Astralago-Festucetum arenariae).

In both districts, only minor areas are in arable cultivation or long ley grassland (Lolio-Cynosuretum). The major portion, however, is limited by soil restrictions resulting from coarse textures and by risk of erosion. Whereas some areas are moderately suited for improved grassland, the remainder affords only rough grazing of moderate grazing values.

THE GLENEAGLES/AUCHENBLAE/COLLIESTON/DARNAWAY ASSOCIATIONS

(Map unit 273)

These four associations have been grouped as all are developed on similar or closely related parent materials, defined as fluvioglacial sands and gravels derived from Old Red Sandstone sediments and lavas, and acid schists. Only one map unit has been distinguished, map unit 273, which covers 144 square kilometres (0.6 per cent of the land area) and is widespread in Eastern Scotland (Sheet 5). It comprises brown forest soils, humus-iron podzols and some gleys. On the Moray Firth Lowlands in the north, it occurs as humus-iron podzols of the Darnaway Association developed on reddish brown sands and gravels derived from Old Red Sandstone sediments and acid schists; on the Grampian Lowlands in the Newburgh area and to the south of Aberdeen, it is found as brown forest soils of the Collieston Association formed on reddish brown water-sorted bands of varying texture overlying sandy gravel; in Strathmore and in north-east Fife it occurs as humus-iron podzols of the Auchenblae Association on reddish sands and gravels which, in the type area, owe their colour to the presence of red marls or mudstones; in lower Strathmore, in east Perthshire and in parts of Fife, it occurs as humus-iron podzols of the Gleneagles Association which vary from brown to reddish brown in colour and, in addition to Old Red Sandstone sediments and acid schists, have lava as a component rock.

The unit occurs on mounds and terraces with gentle and strong slopes in climate regions ranging from warm, moderately dry lowlands with an average annual rainfall of 700 millimetres to fairly warm, moderately dry lowlands and foothills with 1000 millimetres rainfall.

Vegetation ranges from arable and permanent pasture (Lolio-Cynosuretum) to dry Atlantic heather moor (part of Carici binervis-Ericetum cinereae) and herb-rich bent–fescue grassland (part of Achilleo-Festucetum tenuifoliae). Locally common broadleaved woodlands are mainly eastern highland oakwood (Trientali-Betuletum pendulae) and southern oakwood (Galio saxatilis-Quercetum).

In parts of the Moray Firth Lowlands, in much of the North-East Lowlands and in much of the area where it occurs in Strathmore, east Perthshire and Fife, the land is cultivated and produces a moderate range of crops. Coarse texture and slope are the principal limitations. Where the restrictions are more severe, land use is limited to a narrow range of crops based on grassland. The few areas of woodland and moorland are mainly moderately or marginally suited for grassland reclamation.

THE GOURDIE/CALLANDER/STRATHFINELLA ASSOCIATIONS
(Map units 274–277)

The soils of the three associations comprising this group are all developed on drifts derived mainly from acid metamorphic rocks and Lower Old Red Sandstone sediments. Though parent materials of both the Gourdie and Strathfinella Associations also contain a proportion of igneous rocks, and are lithologically similar, they differ in texture and colour. Gourdie parent material varies in texture from loam to fine sandy clay loam and in colour from brown to light yellowish brown. That of the Strathfinella Association is coarser in texture and notably reddish brown in colour. The combined associations cover 350 square kilometres (1.4 per cent of the land area).

The drift forming the parent material of the Callander Association is derived mainly from grey sandstones and slaty metamorphic rocks, although some schistose rocks are present. The texture is normally fine sandy loam and the colour is yellowish brown.

The Strathfinella Association occurs along the slopes of the Grampian Foothills, extending from the outskirts of Stonehaven to near Tullo Hill, north-west of Brechin. The Gourdie Association, most extensive of the three, occurs in a belt of country about 6 kilometres broad following the line of the Highland Boundary Fault from the Kirriemuir district in Angus, south-west to Glen Artney in Perthshire. The Callander Association, which occurs only in central Perthshire, extends from Callander south-east down the valley of the River Teith to Doune and south-west by Aberfoyle to Loch Lomond.

Four map units have been distinguished in climate regions varying from fairly warm, moderately dry lowlands and foothills with an average annual rainfall of 800 millimetres in the north-east, through cool, wet foothills and uplands with 1000 millimetres and fairly warm, wet lowlands and foothills with up to 1400 millimetres to warm, wet lowlands with 1800 millimetres in the south-west.

Map unit 274   The most widespread of the four map units, it covers 202 square kilometres (56 per cent of the associations) and extends along the line of the Highland Boundary Fault from Kirriemuir through Perthshire to Aberfoyle. It consists mainly of freely and imperfectly drained brown forest soils with some humus-iron podzols and gleys and occurs on undulating lowlands and foothills with gentle and strong slopes. Because the combined clay (<2 μm) and silt (2–50 μm) contents of the parent material total around 40 per cent, the soils generally have good moisture- and nutrient-holding capacities. At altitudes below 90 metres, the unit provides good arable land which produces a moderate range of crops. At higher altitudes, or on steeper slopes, land use is restricted to a narrow range of crops based on long ley grassland. Uncultivated ground is moderately or marginally suited for grassland reclamation or is restricted to rough grazings of high or moderate grazing values. Occasional areas of broadleaved woodland consist of eastern highland oakwood and birchwood (Trientali-Betuletum pendulae) and southern oakwood (Galio saxatilis-Quercetum).

Map unit 275   With an area of 63 square kilometres (18 per cent of the associations), the unit comprises the wetter soils of this group of associations and consists mainly of noncalcareous and peaty gleys with some brown forest soils with gleying and peat. Occurring largely on gentle slopes, and in hollows of undulating lowlands and foothills, the unit is found throughout areas of the Gourdie and Callander Associations. The soils are largely uncultivated and support flying bent grassland (part of Junco squarrosi-Festucetum tenuifoliae), sharp-flowered rush pasture (Potentillo-Juncetum acutiflorae), and lowland blanket and flying bent bogs (parts of Erico-Sphagnetum papillosi). The land is moderately or marginally suited for grassland reclamation, though a few areas produce a narrow range of crops based on long ley grassland.

Map unit 276   Covering an area of 85 square kilometres (24 per cent of the associations), it occurs mainly in the area occupied by the Strathfinella Association and, in a few localities, by the Callander Association. The unit is found on undulating foothills and uplands with gentle and strong slopes and consists largely of humus-iron podzols with some brown forest soils, peaty podzols and gleys. The cultivated land produces a moderate or narrow range of crops, the latter based on long ley grassland. Gradient is the main limiting factor to arable agriculture. Uncultivated areas support dry and moist Atlantic heather moors (parts of Carici binervis-Ericetum cinereae), common white bent grassland (part of Junco squarrosi-Festucetum tenuifoliae), sedge mires and rush pastures. Acid bent–fescue grassland (part of Achilleo-Festucetum tenuifoliae) is common on the freely drained brown forest soils and podzols of the footslopes. These are moderately or marginally suited to grassland reclamation or are restricted to rough grazings which have moderate and high grazing values.

Map unit 277   This unit covers only a small area (less than 1 square kilometre) on the hill slopes and valley sides above Callander. It consists of peaty and humus-iron podzols with some gleys and peat and is uncultivated, supporting dry and moist Atlantic heather moors (parts of Carici binervis-Ericetum cinereae), acid bent–fescue grassland (part of Achilleo-Festucetum tenuifoliae) and heath rush–fescue grasslands (Junco squarrosi-Festucetum tenuifoliae). Land use is restricted entirely to rough grazings. Grazing values are low and moderate.

THE HATTON/TOMINTOUL/KESSOCK ASSOCIATIONS
(Map units 281–286)

The soils of these map units are developed on drifts derived from Middle and Lower Old Red Sandstone rocks, dominantly conglomerates which contain pebbles and cobbles of quartzite, schist, granulite or granite or combinations of these rock types. Brown, red and green sandstones and occasional lime-rich sediments contribute locally to the parent material of the soils.

The drifts are stony and mainly reddish brown and range from loamy sands to sandy clay loams. They are usually shallow and coarse textured on convex slopes, summits and ridges; rock outcrops are sometimes present in these situations. Deeper, colluvial drifts mantle some of the steep slopes. On lower and concave slopes, a sandy loam or sandy clay loam till is usual.

On the published 1:63 360 soil maps of Banff (Sheet 96), Huntly (86) and Peterhead and Fraserburgh (87/97), the soils belong to the Hatton Association and are formed on drifts containing many quartzite cobbles. On the Tomintoul map (Sheet 75), soils of the Tomintoul Association have been distinguished on drifts derived from conglomerates and with a greater admixture of sandstones. The Kessock Association has been separated on shallow skeletal drifts derived *in situ* from conglomerate on the published 1:63 360 maps of Cromarty and

Invergordon (Sheet 94), the Black Isle (part of Sheets 83, 84, 93 and 94) and Nairn and Cromarty (Sheet 84 and part of 94). Soils in the other areas are not covered by published maps and have not yet been assigned to a particular association.

The map units cover 232 square kilometres or 0.1 per cent of the land area. The main areas are between Fyvie and New Aberdour, around Tomintoul and near Evanton; smaller areas occur elsewhere in Ross and Cromarty and in Inverness-shire. Landforms include hills and valley sides with gentle to very steep slopes, and undulating lowlands with gentle and strong slopes. The units range from non- to very rocky, the non-rocky areas predominating.

Average annual rainfall ranges from 700 millimetres on the Black Isle to 1600 millimetres on Meall Fuar-Mhonaidh (696 metres) south-west of Drumnadrochit. Climatic regions range from warm, moderately dry lowlands to cold wet uplands, with most areas in the fairly warm, moderately dry lowlands and foothills and cool, wet foothills and uplands. The vegetation includes Atlantic and boreal heather moors (Carici binervis-Ericetum cinereae and Vaccinio-Ericetum cinereae), arable and permanent pastures (Lolio-Cynosuretum) and blanket bog communities (Erico-Sphagnetum papillosi).

As most of the parent materials are acid and low in exchangeable bases, brown forest foils are not extensive. These soils are restricted to lower slopes, to steep slopes where the soils are subject to rejuvenation, and around outcrops of lime-rich sediments which may also be responsible for the presence, in the Tomintoul area, of plants typical of base-rich, flushed soils. The most extensive soils are humus-iron and peaty podzols. Restricted mainly to lower and concave slopes are noncalcareous and peaty gleys. In the Hatton and Tomintoul Associations, gleys are more common than in most neighbouring associations, probably because of the relatively fine texture of the till. Rankers occur around outcrops and where rock is close to the surface. Subalpine soils are limited to the areas above 500 metres altitude.

Map unit 281   Consisting of noncalcareous and peaty gleys with some humic gleys, humus-iron and peaty podzols, peat and brown forest soils it covers 51 square kilometres (22 per cent of the associations) between Fyvie and New Aberdour and around Tomintoul, with a very small area near Evanton. The unit occurs on undulating lowlands and hills with gentle and strong, mainly concave, slopes and is non-rocky.

The soils are developed on greyish brown or reddish brown till which ranges from sandy loam to sandy clay loam. Plant communities include arable and permanent pastures (Lolio-Cynosuretum), rush pastures and sedge mires, and bog heather moors (Narthecio-Ericetum tetralicis). The land is used for arable agriculture, rough grazing and forestry. It has a capability for agriculture which varies from a moderate range of arable crops, between Fyvie and New Aberdour, to a moderate suitability for grassland reclamation in the Tomintoul area on higher ground.

Map unit 282   This map unit comprises humus-iron podzols with some non-calcareous gleys and peaty podzols and rankers. It is the most extensive unit of the association, covering 89 square kilometres (38 per cent of the associations). The main areas are between Fyvie and New Aberdour, and around Tomintoul and Evanton, with smaller areas on the Black Isle, near Loch Ussie and south of Beauly. It occurs on undulating lowlands and hills with gentle to very steep, often convex, slopes. The relief is deeply dissected in the Tore of Troup area near New

Aberdour. Most of the unit is non-rocky, but in the smaller areas listed above, the drift cover is usually thin and rock crops out locally.

Most soils are freely drained humus-iron podzols though iron podzols occur in the Evanton area. Some podzols in the map unit have indurated B horizons. Soils on lower slopes are often cultivated. Noncalcareous and peaty gleys occur in receiving sites, usually concave, lower slopes. Brown forest soils, both freely and imperfectly drained, are restricted mainly to the footslopes of Windyheads Hill, near New Aberdour. Peaty podzols occur on Windyheads Hill and on upper slopes in the Evanton area. There are a few rankers and lithosols on ridge tops, for example, on the Black Isle. These have organic surface layers and up to 10 centimetres of an A horizon overlying hard conglomerate bed-rock.

Plant communities on the podzols include dry and moist boreal heather moors with some Atlantic heather moor (parts of Vaccinio-Ericetum tetralicis and Carici binervis-Ericetum cinereae)and boreal juniper scrub (Trientali-Juniperetum communis), and arable and permanent pastures (Lolio-Cynosuretum) where the soils are cultivated.

The land is used for forestry, rough grazing and arable agriculture. Wooded slopes, north of Evanton, are notable for prolific natural regeneration of a number of tree species, especially western hemlock. The land capability for agriculture has a wide range though arable land capable of producing a moderate or narrow range of crops is the most extensive. The uncultivated moorland is suited for grassland reclamation, but becomes restricted to rough grazings where limitations become more severe.

Map unit 283   Comprising peaty podzols with peaty gleys, peat and humus-iron podzols it covers 7 square kilometres (3 per cent of the associations) and occurs on Windyheads Hill near New Aberdour and in the Tomintoul area. The landforms are non-rocky hills and valley sides with gentle and strong slopes.

The soils are formed on a stony sandy loam or loamy sand drift. On Windyheads Hill the soils often have an indurated B horizon, but in the Tomintoul area induration is only local. Most peaty podzols have a wet peaty surface layer, a gleyed E horizon, and an iron pan.

The vegetation is mainly moist boreal heather moor (part of Vaccinio-Ericetum cinereae), with bog heather moors (Narthecio-Ericetum tetralicis) on peaty gleys, and blanket bogs (Erico-Sphagnetum papillosi) on peat. The land is used for rough grazing, forestry and grouse-shooting. The capability for agriculture is limited mainly to a moderate or marginal suitability for grassland reclamation. Increasing altitude further restricts small areas to rough grazings of low grazing value.

Map unit 284   This map unit consists of peaty podzols and peat with some peaty and noncalcareous gleys, humus-iron podzols and, above 500 metres in the Evanton area, subalpine soils. The unit occupies 29 square kilometres (13 per cent of the associations) and occurs mainly to the north-west of Evanton with a smaller area near Tomintoul. The landforms are rounded hills with gentle to steep slopes. Cnoc Céislein, in the Evanton area, has a number of cols and an extensive valley floor on its western side. Most of the unit is non-rocky, but in the Evanton area, rock, mainly conglomerate, crops out locally and is usually close to the surface on convex slopes.

Most of the peaty podzols have an iron pan, and on lower concave slopes they often have a thick peaty surface layer and a strongly gleyed E horizon. The peat occurs mainly on gentle slopes.

Plant communities include moist Atlantic heather moor (part of Carici binervis-Ericetum cinereae) and boreal heather moors (Vaccinio-Ericetum cinereae) on peaty podzols, blanket bog communities (Erico-Sphagnetum papillosi) on peat, and bog heather moors (Narthecio-Ericetum tetralicis) on peaty gleys. The land is used for rough grazings, sport and forestry, the agricultural capability being restricted to grazings of low value.

Map unit 285   This unit consists of peaty and humus-iron podzols with some rankers, peat and peaty gleys. It covers 27 square kilometres (12 per cent of the associations) and has a scattered distribution from near the Orrin Falls, west of Muir of Ord, to the Loch Duntelchaig area on the eastern side of Loch Ness. The relief is usually distinctive: it has slightly to moderately rocky knolls and hills with smooth, steep, convex slopes and rounded summits. The cols and hollows are drift infilled.

On the knolls and hills, the drift is shallow and stony. Most soils are podzols, with rankers mainly on convex slopes where conglomerate is close to the surface. Peat and peaty gleys occur in cols and hollows.

On the podzols, the vegetation is dry and moist Atlantic heather moors (parts of Carici binervis-Ericetum cinereae), with some acid bent–fescue grassland (part of Achilleo-Festucetum tenuifoliae) on the steepest slopes. The peat and peaty gleys support blanket bogs (Erico-Sphagnetum papillosi) and bog heather moors (Narthecio-Ericetum tetralicis), respectively.

The land is used mainly for rough grazing, with some forestry. Capability for agriculture is limited, generally, to rough grazings of low value though small areas vary from a narrow range of crops based on long ley grassland to a high suitability for grassland reclamation and to rough grazings with moderate grazing values.

Map unit 286   This map unit consists of rankers with some peaty and humus-iron podzols, peat and subalpine podzols. It occupies 29 square kilometres (13 per cent of the associations) mainly on the eastern side of Loch Ness from Loch Duntelchaig to Foyers, with smaller areas on Meall Fuar-Mhonaidh and near Contin. It occurs on very rocky, rugged hills with strong to very steep slopes.

The drift cover is usually very shallow, and peaty and podzolic rankers are dominant, with some peaty and humus-iron podzols where the drift is thicker. Peat occurs on gentler slopes, for example on the northern side of Meall Fuar-Mhonaidh. This hill is capped by subalpine podzols.

The vegetation is mainly dry and moist Atlantic heather moors (parts of Carici binervis-Ericetum cinereae), with some acid bent–fescue grassland (part of Achilleo-Festucetum tenuifoliae) and, on subalpine soils, alpine azalea–lichen heath (Alectorio-Callunetum vulgaris). The land is used for rough grazing and locally for forestry. Most of the rough grazings have only a low grazing value though limited areas have a moderate value.

THE HINDSWARD ASSOCIATION

(Map unit 291)

Soils of Hindsward Association are developed on drifts derived from Carboniferous sediments and basic igneous rocks. In Eastern Scotland (Sheet 5), the association is confined to Fife where it occurs in the vicinity of several of the basalt hills, being most extensive in east Fife. Only one map unit has been separated, map unit 291, which covers an area of 27 square kilometres (0.1 per cent of the land area). It consists of brown forest soils with gleying, occurring on undulating lowlands and foothills with gentle and strong slopes, together with some noncalcareous and humic gleys which occupy hollows. The dominant soil series within the unit has a dark brown loam topsoil overlying a subsoil horizon of grey sandy clay loam with frequent stones, mainly igneous. This latter horizon passes into a grey clay loam drift containing many stones, both igneous and sandstone. The unit occurs in the climate region of fairly warm, moderately dry lowland and foothill with an average annual rainfall ranging from 750 to 950 millimetres. Much of the unit is cultivated and provides a moderate range of crops. A few areas are in permanent pasture (Lolio-Cynosuretum). Fine textures and frequently stoniness are the main limiting factors.

THE INSCH ASSOCIATION

(Map units 316–320, 323, 324, 326, 328–330)

The soils of the Insch Association are developed on drifts derived from gabbros including norite and metamorphosed igneous rocks such as hornblende-schist and epidiorite. On concave slopes and in basins, in the lowlands and foothills, a grey sandy loam to clay loam till with ochreous mottling is common. On convex slopes however, and generally in the uplands and mountains, there is a stonier, yellowish brown sandy loam or loamy sand drift; rock outcrops occur locally. In many of these deposits, chemically weathered material derived from the underlying basic igneous rock has contributed substantially to the solum. Exposures of the weathered rock are particularly common in the Cabrach area.

In Eastern Scotland (Sheet 5), the association is restricted almost entirely to the counties of Aberdeen and Banff. The main areas are the Insch valley, the Cabrach–Strathdon–Morven and Huntly districts. Smaller areas occur near Balmedie and near Portsoy, and there is an outlying area to the north-west of Crieff in Glen Lednock. The association covers 411 square kilometres or 1.7 per cent of the land area and elevations range from 50 to 870 metres. The landforms include undulating lowlands, hills and valley sides, and mountain summits. Slopes are usually gentle in the lowlands, but elsewhere they range up to steep. The map units vary from non- to very rocky, and rocky areas are usually bouldery.

The average annual rainfall ranges from 800 to 1200 millimetres, except for the outlying area in Glen Lednock where it reaches 1600 millimetres. The climatic regions range from fairly warm, moderately dry lowland and foothill, which is the most extensive category, to very cold, wet mountain.

Plant communities include arable and permanent pastures (Lolio-Cynosuretum), rush pastures, dry and herb-rich boreal heather moors (parts of Vaccinio-Ericetum cinereae), bent–fescue grasslands (Achilleo-Festucetum tenuifoliae), upland and mountain blanket bogs (part of Erico-Sphagnetum papillosi and Rhytidiadelpho-Sphagnetum fusci respectively), and alpine azalea–lichen heath (Alectorio-Callunetum vulgaris).

The soils are mostly brown forest soils and humus-iron podzols. Because of their higher base status inherited from the parent materials, brown forest soils reach higher elevations than those of more acid, neighbouring associations.

Humus-iron podzols differ from those on more acid parent materials by lacking a clear, ash-coloured E horizon below the organic surface layers. This is due to the scarcity of quartz in the basic igneous rocks. Peaty podzols are rela-

Plate 13. *The Insch valley, showing good arable land (Class 3.1) on the base-rich brown forest soils of map unit 316 (Insch Association). In contrast, Bennachie in the middle distance is a granite hill with peaty podzols of map unit 117 (Countesswells Association, Classes 5 and 6).* Cambridge University Collection: copyright reserved.

tively uncommon, again probably a consequence of the less acid nature of the parent material. Most brown forest soils and humus-iron podzols have an indurated B horizon. The induration is weak, however, and is not as hard as, for example, in soils of the Countesswells Association. Imperfectly and poorly drained soils usually have prominent, ochreous mottles in the subsoil due to abundant iron from the ferromagnesian minerals. Most alpine and subalpine soils are freely drained. Organic-rich subsoils of alpine soils may have a higher organic-matter content than is general on more acid parent materials. Because of the predominance of basic minerals, adequate amounts of copper and cobalt are usually present. Amounts of total phosphorus are high, but nevertheless trees often respond to applications of phosphorus, suggesting that this element may be non-available.

Map unit 316 This map unit consists of brown forest soils, both freely and imperfectly drained, with some humus-iron podzols and noncalcareous gleys. It covers 257 square kilometres (63 per cent of the association) mainly in the Insch valley (Plate 13). There are smaller areas near Rhynie, Balmedie, Huntly, Strathdon, Logie Coldstone and Portsoy, and in the Cabrach and Glen Lednock areas. The unit occurs on undulating lowlands with gentle slopes and on rounded foothills with gentle and strong slopes; it is non-rocky.

119

The topsoils of the brown forest soils usually have a well-developed crumb structure and are a rich dark brown. Humus-iron podzols occur at high elevations, mostly above 350 metres. Gleys are normally confined to receiving sites.

Plant communities include arable and permanent pastures (Lolio-Cynosuretum) with some bent–fescue grasslands (Achilleo-Festucetum tenuifoliae) and herb-rich and dry boreal heather moors (parts of Vaccinio-Ericetum cinereae). The herb-rich boreal heather moor is associated with brown forest soils that have very dark brown organic-rich A horizons.

Cultivated ground produces mainly a moderate range of crops of high or average yields. Some areas, however, are capable of producing a wider range, whereas other areas are restricted to a narrow range based primarily on long ley grassland. Many of the latter areas, particularly in the Cabrach and Strathdon districts, have excellent soils but are restricted in land use because of the climate. Uncultivated ground is usually moderately suited to grassland reclamation.

Map unit 317   This unit comprises noncalcareous and peaty gleys, with some brown forest soils, both freely and imperfectly drained humus-iron podzols, peat and humic gleys. It covers 55 square kilometres (13 per cent of the association), the main areas being in the Insch valley, around Huntly and in the Cabrach–Strathdon–Morven district. The unit occurs on lower, concave gentle slopes and in basins and valley floors. It is non-rocky.

Most of the soils are associated with the grey sandy loam to clay loam till, the upper part of which is occasionally coarser in texture due to the sorting action of meltwater.

The main plant communities are arable and permanent pastures (Lolio-Cynosuretum) and rush pastures. Many others occur, including sedge mires, white bent–tussock-grass grassland (the *Cirsium palustre–Nardus stricta* Community) and bog heather moors (Narthecio-Ericetum tetralicis) on peaty gleys.

In the Insch valley, most land is under arable cultivation and produces a moderate range of crops, whereas around Huntly it is mainly restricted to long ley pasture. Within the Cabrach–Strathdon–Morven district, most of the soils are peaty gleys at elevations of over 300 metres and the land is only marginally suited to grassland reclamation.

Map unit 318   The soils of this unit are humus-iron podzols with some brown forest soils, noncalcareous and peaty gleys, and peaty podzols. The unit occupies 76 square kilometres (18 per cent of the association), the main area being the Cabrach–Strathdon–Morven district. It occurs on rounded hills and valley sides with gentle to steep slopes. Although mainly non-rocky, tors are present locally on the summits, for example, in the Cabrach area.

The humus-iron podzols usually lack a clear podzolic E horizon, although a whitish coloration develops on a dried-out exposed surface. These soils tend to have Bs horizons of lower chroma, usually 4 to 5, than those of the more acid associations.

The dominant vegetation is dry boreal heather moor (part of Vaccinio-Ericetum cinereae) which is associated with the predominant humus-iron podzols. Other plant communities include heath rush–fescue grasslands (Junco squarrosi-Festucetum tenuifoliae) and acid bent–fescue grassland (part of Achilleo-Festucetum tenuifoliae), arable and permanent pastures (Lolio-Cynosuretum) and boreal juniper scrub (Trientali-Juniperetum communis).

The land is used mainly for rough grazing, grouse-shooting and forestry but much of it is moderately suited to improved grazing. Steepness of slope is the dominant limiting factor.

Map unit 319   This unit consists mainly of peaty podzols, with some peat, peaty gleys and peaty rankers. It occupies only 1 square kilometre (less than 1 per cent of the association) in one area on the northern slopes of Morven. The unit occurs above about 500 metres elevation on gentle to steep slopes and is locally slightly rocky.

The plant communities include dry and moist boreal heather moors (parts of Vaccinio-Ericetum cinereae) on peaty podzols, bog heather moors (Narthecio-Ericetum tetralicis) on peaty gleys, and lowland, upland and mountain blanket bogs (parts of Erico-Sphagnetum papillosi and Rhytidiadelpho-Sphagnetum fusci) on peat. The land is used for rough grazing and grouse-shooting. The grazing values are low.

Map unit 320   This map unit consists of peaty podzols and peat, with some humus-iron podzols, peaty gleys and peaty rankers. It occupies 6 square kilometres (2 per cent of the association) in the Strathdon–Morven area. The unit occurs mainly on rounded hills with gentle and strong, non-rocky slopes though locally slopes are steep and some are slightly rocky.

Most of the peaty podzols and peat are on gentle upper slopes and on plateaux, whereas the humus-iron podzols and peaty rankers occur on gentle to steep, lower slopes.

The vegetation is dry and moist boreal heather moors (parts of Vaccinio-Ericetum cinereae) on the podzols and peaty rankers, upland and mountain blanket bogs (part of Erico-Sphagnetum papillosi and Rhytidiadelpho-Sphagnetum fusci) on peat and bog heather moors (Narthecio-Ericetum tetralicis) on peaty gleys. The land is used for rough grazing and grouse-shooting. Agriculture is restricted largely to the rough grazings which are mostly of low grazing value. Some areas provide only very poor and seasonal grazing. Other limited areas are moderately suited to grassland reclamation.

Map unit 323   The soils of this unit are brown forest soils and brown rankers, with some humus-iron podzols and gleys. Occupying 7 square kilometres (2 per cent of the association) in Glen Gairn near Ballater and in Glen Lednock, the unit occurs on hills and valley sides with strong to very steep slopes. The unit is slightly or moderately rocky, and some of the steeper slopes are scree-covered.

The soils are formed on very stony drifts, with rock usually close to the surface. The plant communities include herb-rich boreal heather moor (part of Vaccinio-Ericetum cinereae), bent–fescue grasslands (Achilleo-Festucetum tenuifoliae), boreal juniper scrub (Trientali-Juniperetum communis) and eastern highland birchwood (part of Trientali-Betuletum pendulae). The land is used for rough grazing. Slope and stoniness are the dominant limiting factors and grazing values are low.

Map unit 324   Comprising humus-iron podzols with some peaty gleys and rankers it occupies 1 square kilometre (less than 1 per cent of the association) in the Cabrach area, and occurs on landforms which are similar to those of *map unit 323*, except that very steep slopes and scree are absent.

The map unit is similar to *map unit 323* but podzols replace the brown forest soils and peaty rankers replace the brown rankers.

The vegetation is mainly dry and moist boreal heather moors (parts of Vaccinio-Ericetum cinereae). The land is used for rough grazing but is moderately suited to grassland reclamation.

Map unit 326 The soils of this unit are rankers and humus-iron podzols, with some brown forest soils. The unit occupies only 1 square kilometre (less than 1 per cent of the association) in an area to the north-east of Morven. It occurs on rocky hills with strong to very steep slopes, some of which are scree-covered. The unit is similar to *map unit 323* but is rockier and has mainly podzolic soils.

The vegetation is mainly dry boreal heather moor (part of Vaccinium-Ericetum cinereae). The land is used for rough grazing and grouse-shooting, and has a limited capability for agriculture. Grazing values are low.

Map unit 328 This unit has subalpine soils, mainly podzols, and peaty podzols with some shallow peat. It occupies 2 square kilometres in Glen Lednock (less than 1 per cent of the association). It occurs on smooth, gentle, upper hill slopes above 600 metres elevation and is non- to slightly rocky.

The soils are formed on a drift derived from a basic diorite.

The vegetation includes boreal heather moors (parts of Vaccinium-Ericetum cinereae) alpine azalea–lichen heath (Alectorio-Callunetum vulgaris) on the subalpine soils, and mountain blanket bog (Rhytidiadelpho-Sphagnetum fusci) on peat. The land is used for rough grazing. Grazing values are low.

Map unit 329 Consisting of peat and alpine soils with some subalpine soils and occasional peaty rankers it occupies 2 square kilometres (less than 1 per cent of the association) near the hill of Morven. The unit occurs above 600 metres elevation on rounded hill summits and upper, gentle and strong slopes, and is mainly non-rocky, locally slightly rocky.

The alpine soils, mostly podzols, occur on summits and convex slopes, and the peat in cols and on gentle straight and concave slopes.

The vegetation includes alpine azalea–lichen heath (Alectorio-Callunetum vulgaris) on alpine soils and mountain blanket bog (Rhytidiadelpho-Sphagnetum fusci) on peat. The land is used for rough grazing and grouse-shooting. The main limitations, altitude and exposure, restrict agriculture mainly to seasonal and very poor grazings.

Map unit 330 The soils of this unit are mainly alpine soils, with some rankers, subalpine soils and peat. It occupies 3 square kilometres (less than 1 per cent of the association) on the hill of Morven (870 metres) and has gentle to steep slopes. It is mainly a slightly rocky unit, with runs of scree on steep slopes.

The drift is usually very stony. Pockets of less stony, yellowish brown, loamy sand drift occur in the col near the summit. The rankers are found where rock is close to the surface and on stabilized scree.

The vegetation is mainly alpine azalea–lichen heath (Alectorio-Callunetum vulgaris). The land is used for rough grazing and ptarmigan-shooting. Altitude and exposure limit agriculture to very poor grazing of a seasonal nature.

THE KIPPEN/LARGS ASSOCIATIONS

(Map units 337–339, 341, 343 and 344)

The soils of this group are developed on drifts derived mainly from sandstones of Upper Old Red Sandstone age. Only the Kippen Association is found in Eastern

Scotland (Sheet 5), where it covers 95 square kilometres (0.4 per cent of the land area). South of the River Tay, it has been mapped in several localities along a general line running south-west from Cupar in Fife through Kinross to Kippen village in the upper Forth valley. A small area occurs north of the Tay on the western outskirts of Dundee.

Six map units have been distinguished, occurring in climate regions ranging from warm, moderately dry lowlands with an average annual rainfall of 760 millimetres in the east, through fairly warm, moderately dry lowlands and foothills with 1000 millimetres rainfall, to fairly warm, wet lowlands and foothills with 1300 millimetres in the west.

Map unit 337   This the largest unit in the association and covers 50 square kilometres (55 per cent of the association). It comprises freely and imperfectly drained brown forest soils with some gleys and is found throughout the whole range of the association, the more extensive areas occurring to the south of Cupar, to the north and north-east of Kinross and to the east of Kippen village. The major component of the unit, the brown forest soil, has a loam topsoil on a reddish brown sandy clay loam with a subangular blocky structure overlying a reddish brown clay loam which has a coarse angular blocky structure. Much of the unit is in arable cultivation or in permanent bent–fescue grassland (Achilleo-Festucetum tenuifoliae). A few small areas support southern oakwood and birchwoods (Galio saxatilis-Quercetum). The more easterly arable areas produce a wide range of crops with high yields; further west, where climate is less favourable and topography is an added limitation, the arable rotations become less flexible with grass husbandry becoming more important.

Map unit 338   The unit is not extensive (7 square kilometres, 10 per cent of the association) and consists mainly of noncalcareous gleys with some brown forest soils with gleying. It occurs in a climate region of fairly warm, wet lowlands and foothills and the moderately severe limitations restrict the land use to long ley grassland. Uncultivated areas, frequently supporting rush pastures and sedge mires, are moderately or marginally suited to grassland reclamation.

Map unit 339   This is the second largest unit in the association, covering 22 square kilometres (25 per cent of the association) and occurring mainly on lower and middle slopes of East Lomond, West Lomond and Benarty Hills. It consists of brown forest soils, largely freely drained, with some humus-iron podzols. Lower slopes are in arable cultivation or in acid bent–fescue grassland (part of Achilleo-Festucetum tenuifoliae). The arable areas produce a moderate or narrow range of crops, the latter range based on long ley grassland. Upper slopes support southern oakwood and birchwoods (Galio saxatilis-Quercetum) or dry Atlantic heather moor (part of Carici binervis-Ericetum cinereae). These slopes are moderately or marginally suited to grassland reclamation. Some areas are restricted to rough grazings which have moderate grazing values.

Map unit 341   This unit covers 7 square kilometres (10 per cent of the association) mainly on upper west-facing slopes of West Lomond Hill and Bishop's Hill and in smaller areas to the west and south-west of Kippen village. It occurs over an altitude range from 180 to 420 metres and consists largely of humus-iron podzols with some brown forest soils and gleys. Although small areas on the lower slopes of West Lomond Hill and to the south and south-east of Kippen support long ley grassland, most of the unit occurs on steep slopes and valley sides and land use is restricted to rough grazings. Grazing values are low. These slopes support dry Atlantic heather moor (part of Carici binervis-Ericetum cinereae), common white bent grassland (part of Junco squarrosi-Festucetum tenuifoliae) and eastern highland birchwood (part of Trientali-Betuletum pendulae).

Map unit 343 Covering only 3 square kilometres (less than 5 per cent of the association) map unit 343 occurs on low hills and valley sides to the south-west of Kippen and comprises peaty podzols and peat with some peaty rankers and minor gleys. Vegetation ranges from dry and moist Atlantic heather moors (parts of Carici binervis-Ericetum cinereae) and common white bent grassland (part of Junco squarrosi-Festucetum tenuifoliae) to bog heather moors (Narthecio-Ericetum tetralicis) and blanket bogs (Erico-Sphagnetum papillosi). The land is moderately suited to grassland reclamation.

Map unit 344 This unit consists of peaty gleys and peat with some peaty rankers and is found contiguous with *map unit 343* on undulating lowlands and foothills. It covers 6 square kilometres (5 per cent of the association). Plant communities include moist Atlantic heather moor (part of Carici binervis-Ericetum cinereae), bog heather moors (Narthecio-Ericetum tetralicis), flying bent bog (part of Erico-Sphagnetum papillosi) and arable and permanent, mainly wet, pastures (Lolio-Cynosuretum). The land is marginally suited to grassland reclamation.

THE LAURENCEKIRK ASSOCIATION

(Map unit 368)

The soils of the Laurencekirk Association are developed on drifts derived from Lower Old Red Sandstone marls and mudstones. The drifts, which vary in colour from reddish brown to weak-red and in texture from sandy clay loam to clay loam, contain a proportion of angular pieces of soft red marl. In several localities, the drift is absent or very thin and the soil parent material consists almost entirely of decomposed marl rock.

Covering a total area of 91 square kilometres (0.4 per cent of the land area) the association occurs most extensively round Laurencekirk in the Howe of the Mearns; smaller areas occur to the north and west of Perth in lower Strathmore and near Crieff and Aberfoyle in west Perthshire.

Map unit 368 is the only map unit distinguished. It consists of imperfectly drained brown forest soils with gleying, together with some freely drained brown forest soils and noncalcareous gleys. The dominant component has a loam topsoil overlying clay loam subsoil horizons. Bordering river and stream channels in a few limited areas, the soils have undergone modification due mainly to water-sorting. Their upper horizons are coarse-textured sandy loams or stony gritty loams, sometimes to a depth of 50 centimetres or more, as in the Forfar Association. The underlying drift is generally a red sandy clay loam with a massive structure.

The association occurs in climate regions ranging from warm moderately dry lowlands and foothills with an average annual rainfall around 800 millimetres, through warm, moderately dry lowlands with 760 millimetres rainfall to fairly warm, wet lowlands and foothills with 1300 millimetres rainfall. The greater part of the unit is arable and much of it is highly productive with a wide range of crops. The remaining arable areas produce a moderate range of crops, the

limitations being topography, texture and, occasionally, wetness. A few uncultivated areas are in permanent pasture (Lolio-Cynosuretum) or broadleaved
woodland, mainly elmwood (Querco-Ulmetum glabrae) and eastern highland
oak and birchwood (Trientali-Betuletum pendulae). In the far west, due to
increased rainfall and steeper slopes, the land use is restricted to long ley grassland.

THE LESLIE ASSOCIATION

(Map units 369–371)

The Leslie Association comprises soils that are developed on drifts derived from
ultrabasic and locally some basic rocks. The rocks include serpentinite, troctolite,
olivine-gabbro and picrite. The drifts are brown and grey, stony, sandy loams
and loams. Where rock is close to the surface, the drift is very stony, with
abundant angular stones. Tills, often containing erratic rock fragments such as
granite, schist and gabbro, occur in the lowlands and on concave slopes.

The association is restricted to Aberdeenshire, where the main areas are near
Huntly, on the Green Hill near Strathdon, on the Hill of Towanreef near
Lumsden, Beauty Hill near Newmachar and the Coyles of Muick south of
Ballater. It covers 29 square kilometres, or 0.1 per cent of the land area.
Elevations range from 100 to 600 metres. The landforms include undulating
lowlands, valley sides and hills; slopes are gentle and strong, and locally steep.
Most of the land is non-rocky but rock outcrops are present locally, especially on
summits.

The average annual rainfall is 900 to 1200 millimetres and the climatic regions
range from fairly warm, wet lowlands and foothills to cold, wet uplands. The
plant communities include herb-rich bent-fescue grassland (part of Achilleo-
Festucetum tenuifoliae), arable and permanent pastures (Lolio-Cynosuretum),
boreal juniper scrub, (Trientali-Juniperetum communis), serpentine tussock-
grass grassland (the *Helictotrichon pratense–Deschampsia caespitosa* Association) and sedge mires.

The soils are freely or imperfectly drained brown magnesian soils, magnesian
gleys, and rankers around rock outcrops. Uncultivated soils often have an organic
surface horizon. The ratios of exchangeable magnesium to calcium range from
2:1 to more than 10:1. This contrasts with the ratios of at least 1:5, typical of
non-magnesian soils, that is, with calcium exceeding magnesium. The base
saturation of the B and C horizons is 90 to 100 per cent. The soils on Beauty Hill
have high contents of extractable nickel, high enough to restrict seriously the
growth of cereals unless the surface pH is raised to around 7. Trees often grow
poorly, for reasons which may be associated with phosphorus deficiency, high pH
values, high magnesium to calcium ratios, and toxicities due to high levels of
aluminium and trace elements.

Map unit 369   This unit consists of brown magnesian soils and some magnesian
gleys. It covers 11 square kilometres (40 per cent of the association) on the Green
Hill, the Hill of Towanreef, Beauty Hill and south-west of Huntly and Insch. The
unit occurs on undulating, gently sloping lowlands, and on hills with gentle and
strong slopes. The land is mainly non-rocky, though there are some rock outcrops
on hill summits.

The brown magnesian soils are freely drained on ridges and convex slopes.
Lower and concave slopes often have a hummocky microrelief in which
imperfectly and freely drained soils on mounds are interspersed with gleys in

small flushed channels and hollows. Cultivated soils have a dark-coloured humose A horizon, but the commoner, uncultivated soils have up to 15 centimetres of black organic surface horizon. The freely drained soils have yellowish to reddish brown B horizons whereas in imperfectly drained soils these horizons are greyish with ochreous mottles.

The vegetation includes herb-rich bent–fescue grassland (part of Achilleo-Festucetum tenuifoliae), arable and permanent pastures (Lolio-Cynosuretum), and boreal juniper scrub (Trientali-Juniperetum communis), with sedge mires on magnesian gleys. Herb-rich boreal heather moor (part of Vaccinio-Ericetum cinereae) occurs on freely drained brown magnesian soils on the summit of Green Hill. Near Insch, the land use is mainly arable agriculture capable of a moderate range of crops with high or moderate yields. On Beauty Hill, the agriculture is based on long ley grassland with a narrow range of crops. Elsewhere, the land is moderately suited for grassland reclamation, becoming restricted to rough grazings of moderate grazing value above 450 metres elevation.

Map unit 370 The unit comprises magnesian gleys with some freely or imperfectly drained magnesian soils. The area is 15 square kilometres (50 per cent of the association) mainly in the Bin Wood and Dunbennan Hill near Huntly, with smaller areas on Green Hill, the Hill of Towanreef, in the Cabrach district and in Glenkindie. The unit occurs on undulating lowlands and on the lower concave slopes of hills. Slopes are gentle and strong and the land is non-rocky. Most soils are uncultivated and have a dark-coloured humose or a black organic surface horizon up to 15 centimetres thick. The B horizons are dominated by grey colours and sometimes have a prismatic structure, whereas the C horizon is massive.

The vegetation includes serpentine tussock-grass grassland (the *Helictotrichon pratense–Deschampsia caespitosa* Association), sedge mires and, on the brown magnesian soils, boreal juniper scrub (Trientali-Juniperetum communis) and herb-rich boreal heather moor (part of Vaccinio-Ericetum cinereae). In general, the land is moderately suitable for grassland reclamation. Some limited areas include arable agriculture with a moderate or narrow range of crops and rough grazings of moderate grazing value.

Map unit 371 This unit has brown magnesian soils with some magnesian gleys, rankers and peat. It covers 3 square kilometres (10 per cent of the association) on the Coyles of Muick and on the Green Hill. Though similar to *map unit 369*, it also occurs on a wider range of slope, up to 25 degrees.

The rankers occur around rock outcrops. Most have organic surface horizons on shattered, partly weathered serpentinite.

The plant communities include herb-rich bent–fescue grassland, (part of Achilleo-Festucetum tenuifoliae) and herb-rich boreal heather moor (part of Vaccinio-Ericetum cinereae), with sedge mires on the magnesian gleys.

The land is used mainly for rough grazing which has a moderate grazing value. Rock outcrops and the climate are the dominant limitations.

THE LINKS ASSOCIATION

(Map units 380 and 382)

Links soils are developed on windblown sand, largely quartzose, which overlies raised beach deposits. Although the dominant drainage class is free, there are frequent small areas of imperfect drainage. In certain localities where drainage is poor or very poor, there is often an underlying horizon of fine-textured material. In some areas, for example, between Banff and Portsoy, in the vicinity of Fraserburgh and on the north-east coast of Tentsmuir in Fife, limited deposits of shelly sand are found in the soil profile.

Links soils cover an area of 142 square kilometres or 0.6 per cent of the land area. The most northerly area delineated on the map occurs on the shores of Nigg Bay. The next occurrence is at Culbin Forest where the soils follow the Moray coastline to the estuary of the River Findhorn. From the eastern shores of the Findhorn estuary, soils of the Nigg/Preston Associations developed on un-differentiated raised beach deposits extend round the greater part of Burghead Bay. From there, Links soils follow the coastline almost continuously to Lossiemouth. A small area occurs at Whitehills between Portsoy and Banff, and thereafter, the pattern continues from Fraserburgh to Peterhead and from Collieston to Aberdeen. In Angus, they extend from St Cyrus to Montrose and, further south, they form the Barry Links between Carnoustie and Monifieth. In Fife, they cover Tentsmuir between Tayport and the Eden estuary, Pilmore Links at St Andrews and a small area at Elie. Links soils are found in the climate regions of warm, moderately dry lowlands with an average annual rainfall range of 650 to 750 millimetres apart from the coastal strip from Fraserburgh to Aberdeen which is a region of fairly warm, moderately dry lowlands and foothills with average annual rainfall range of 700 to 760 millimetres.

Two map units have been distinguished in Eastern Scotland (Sheet 5).

Map unit 380   This, the more extensive unit, covers an area of 126 square kilometres (90 per cent of the association) and consists mainly of noncalcareous regosols with some gleys and minor areas of brown forest soils, occurring on beaches and dunes with gentle and strong slopes at altitudes from 5 to 8 metres. The dunes are moundy and unstable and merge seawards with the tidal sand. They are partially fixed by marram grass with lesser amounts of lyme-grass. The links on the landward side are gently sloping to flat and are stabilized under a comparatively close sward of fine-leaved grasses or heath. The main natural plant communities are northern dunes (Elymo-Agropyretum boreo-atlanticum and Elymo-Ammophiletum) and dune pastures (Astragalo-Festucetum arenariae). Under coniferous woodland or under semi-natural vegetation, podzolic profiles have developed. In certain areas, particularly where there are shelly sands, brown forest soils have developed and a limited amount of arable agriculture is possible. Regular ploughing should be avoided because of the danger of wind erosion associated with the poorly developed structure although some areas of Links soils are cropped once in four years and then sown to grass. Because of poor water-and nutrient-holding capacity resulting from coarse texture, and the consequent risk of wind erosion, limitations range from moderately to extremely severe. Consequently, much of the area is restricted to rough grazing. Limited areas, however, for example, near Carnoustie and St Andrews, are cultivated and capable of producing a narrow or moderate range of crops. Elsewhere there is some potential for grassland reclamation. In the stabilized dune system at the south-east section of Barry Links, the agricultural capability is minimal.

Map unit 382   Covering 16 square kilometres (10 per cent of the association), the unit consists mainly of noncalcareous and peaty gleys with some noncal-careous regosols. It occurs on beaches and dune slacks where the topography is level or gently sloping, especially where underlying fine-textured deposits inhibit good drainage. In Aberdeenshire, there are several small areas, one of which, to the north of Peterhead, includes peat and is underlain for the most part by a marine silty clay. In north-east Fife, two areas have been separated, the larger of which, lying to the north of Leuchars, includes soils developed mainly on fine sands and silts and bordering the windblown sands of Tentsmuir.

Because the drainage of the soils is poor in addition to the limitations imposed by coarse texture and lack of structure, land use is restricted mainly to rough grazing. Only limited areas are suitable for reclamation to grassland. Two of the areas listed above, however, have been cultivated and produce a moderate range of crops with average yields. The soils usually support permanent and rush pastures (Lolio-Cynosuretum) or silverweed pasture (the *Potentilla anserina–Carex nigra* Community) and sedge mires.

THE MILLBUIE ASSOCIATION
(Map units 405 and 406)

The Millbuie Association occurs only in the Black Isle where it forms the most extensive association, occupying 199 square kilometres (0.8 per cent of the land area). It stretches from Muir of Ord in the west to Udale Bay and Rosemarkie in the east.

The parent material is a morainic drift derived from sediments of Middle Old Red Sandstone age, mostly brown or yellowish brown sandstones with occasional shales; the shales are sometimes calcareous. The colour of the drift, although usually brown and occasionally yellowish brown at depth, reflects the colour of the sandstones. For example, around Avoch, Kessock and Redcastle and overlying red sandstones and conglomerates, the parent material is reddish brown. North-east of Muir of Ord, this drift contains an abundance of schists. In general, the morainic drift overlies a compact reddish brown till and is thickest in the west. As the morainic drift overlay becomes thinner towards the central and eastern parts of the Black Isle, the underlying till forms a greater part of the soil profile and there is a transition from the Millbuie to the Cromarty Association.

The landforms comprise non-rocky, undulating lowlands with gentle and strong slopes. Throughout much of the unit, the slopes are long and regular and in places rise gently from sea level to almost 200 metres on the ridge which forms the backbone of the Black Isle. Near Munlochy, however, the slopes are variable and around Culbokie and to the east of Conon Bridge there are moundy areas.

Climate regions vary from a coastal belt of warm dry lowlands to fairly warm moderately dry lowlands; the latter applies to most of the association area. The average annual rainfall ranges from 700 to over 800 millimetres.

Humus-iron podzols, freely and imperfectly drained, are the dominant soils in the association. Induration is widespread. Minor and local soils include non-calcareous gleys, peaty gleys and peat; they are mainly associated with depressions.

Map unit 405 This unit of 12 square kilometres (5 per cent of the association) occupies a large area in the western end of the peninsula together with a number of mainly small patches scattered throughout the association.

The large area consists of a kettlehole complex north-west of Tore. On the mounds are humus-iron podzols, in some places tending to peaty podzols with an iron pan. Peaty gleys and peat with limited noncalcareous gleys dominate the hollows and adjacent lower slopes. Bog heather moors (Narthecio-Ericetum tetralicis) and lowland blanket bog (part of Erico-Sphagnetum papillosi) are associated with the wet hollows whereas Atlantic heather moors (Carici binervis-Ericetum cinereae) are related to the podzols on the mounds. Much of the complex is now afforested.

The soils in the scattered areas are mostly noncalcareous gleys with minor peaty gleys; also included are poorly drained soils developed on alluvium and solifluction deposits. These areas are usually restricted to elongated hollows and receiving sites which lie between ridges orientated south-west to north-east. An exception to this general distribution is an area south-west of Mount Eagle where the noncalcarous gleys occupy the broad part of the central spine of the Black Isle. Gentle slopes and shallow induration make lateral drainage difficult. Most of the soils are cultivated and capable of producing a narrow range of crops based on grassland. Wetness is the main limiting factor. The principal plant communities are arable and permanent pastures (Lolio-Cynosuretum) and rush and sedge mires.

Map unit 406   Of the two units in the association this is by far the larger, occupying 187 square kilometres (or 95 per cent of the association).

Humus-iron podzols, both freely and imperfectly drained, dominate the unit; minor noncalcareous gleys are restricted to localized hollows. The imperfectly drained podzols are the most extensive soils. They are developed on gently sloping ground where compact till, or an indurated horizon, impedes soil drainage. Freely drained podzols are common at the western end of the peninsula where they have developed readily on the deeper morainic drift. Eastwards, the coarse drift becomes shallower and freely drained soils are related more to shedding sites and steeper slopes. Towards the top of the central ridge, east of Mount Eagle, some of the podzols have developed an iron pan but lack a well-developed peaty surface horizon. An indurated B horizon, prevalent in both the freely and imperfectly drained, uncultivated podzols, occurs often within 25 centimetres of the surface. In cultivated soils, however, this horizon often lies below 60 centimetres, usually where the topsoil has been selectively deepened.

Most land below 150 metres is under rotational arable agriculture. Above, there is some semi-permanent pasture although much of the central ridge is afforested with a few localized areas of moorland. With increasing altitude there is a gradual deterioration of climate and the soils become shallower and stonier.

Plant communities are dominantly arable and permanent pastures (Lolio-Cynosuretum) with dry and moist Atlantic heather moors (parts of Carici binervis-Ericetum cinereae) on the upper slopes.

Although much of the arable land is capable of producing only a moderate range of crops, limited areas in sheltered coastal sites, for example, near Munlochy and east of Balblair, are capable of producing a wide range of crops with high yields.

THE MOUNTBOY ASSOCIATION
(Map unit 414)

The soils of the Mountboy Association, covering a total area of 214 square kilometres (0.9 per cent of the land area), are formed on drifts derived mainly from Old Red Sandstone lavas and sandstones. The drifts, which vary in texture from loam to sandy clay loam and in colour from reddish brown to weak red, resemble those of the Balrownie Association. Frequently, however, the lava imparts a purplish or brownish hue. In some areas, particularly in the vicinity of igneous intrusions, a coarse-textured drift derived largely from Old Red Sandstone sediments overlies lava rock. When the association was first identified in north-east Angus, the parent material was defined as till derived from Lower Old Red Sandstone lava and sediments. In districts further south, rocks of Upper Old Red Sandstone age are also present and the original parent material definition has been widened to include these.

The most northerly area of the association, amounting to some 60 square kilometres, is located in the coastal belt between Stonehaven and Lunan Bay and includes the upper parts of the Garvock Hills. Further south, the association occurs among the Sidlaw Hills bordering Strathmore and around lava outcrops in the Dundee district. South of the River Tay, in Fife, it is found on middle and lower slopes of the eastern Ochil Hills.

Map unit 414 is the only map unit distinguished in Eastern Scotland (Sheet 5). It comprises brown forest soils with gleying and freely drained brown forest soils, and occurs on undulating lowlands and foothills with gentle and strong slopes. The unit also includes some gleys, mainly in hollows. The dominant soil has a dark brown or dark reddish brown fine sandy loam or loam overlying a reddish brown sandy loam or loam subsoil on a gritty loam or sandy clay loam. Imperfect or poor drainage conditions are associated with these finer textures.

The unit occurs within climate regions ranging from fairly warm moderately dry lowlands and foothills with an average annual rainfall of 760 millimetres in the Garvock Hills district, through warm, moderately dry lowlands with a rainfall 700 millimetres near the coasts of Angus and Fife, to cool, wet foothills and uplands, rainfall 900 millimetres, on parts of the Sidlaw Hills and fairly warm, wet foothills, rainfall 1200 millimetres, on the Ochil Hills.

Much of the unit is under arable cultivation and capable of producing a moderate range of crops, particularly in those areas where there is a high proportion of imperfectly drained brown forest soils. South and south-west of Montrose, on the lower slopes of the Sidlaw Hills, some land is especially productive with a wide range of crops. On upper slopes of many of the hills throughout the unit area where there is a greater proportion of freely drained brown forest soils, the land is uncultivated and is in permanent pasture (Lolio-Cynosuretum). It is highly or moderately suitable for grassland reclamation. Small areas of poorly or very poorly drained gleys, occurring in depressions and hollows, support rush pastures and sedge mires.

THE NIGG/PRESTON ASSOCIATIONS

(Map units 420 and 421)

Due to their similar nature and derivation these two associations have been amalgamated. Their soil parent materials comprise raised beach deposits and marine and estuarine alluvium. Both the Nigg and Preston units are new associations. The former was created to accommodate raised beach deposits which occur in Easter Ross, the Black Isle and along the Moray Firth and which on the published 1:63 360 soil maps had been assigned to a group of miscellaneous beach and alluvial soils; the Preston Association has been mapped only on the 1:250 000 soil map of South-West Scotland (Sheet 7) and is located on a small coastal strip south of Dumfries.

Because of the differing sedimentation conditions, textures within this association vary considerably and range from gravelly sand to silt. Windblown deposits, with fine sand textures, which often accompany littoral deposits, are restricted in distribution.

The Nigg association occupies only 119 square kilometres or 0.5 per cent of the land area. Level or gentle slopes are characteristic, although short steeper slopes, which represent the marine margin of old beach levels, are also present. Occupying a warm, moderately dry lowland climate region, the soils vary in their morphology. Those developed on aeolian sand lack profile development because soil-forming processes have had insufficient time to alter such recent deposits. However, on the highest beach terraces, humus-iron podzols are extensive. Other soils include humus podzols and iron podzols. Poorly drained soils are usually noncalcareous gleys, although where shells are present within the sand, calcareous gleys occur.

The freely and imperfectly drained soils are covered mainly by arable and permanent pastures (Lolio-Cynosuretum) with some dry Atlantic heather moor (part of Carici binervis-Ericetum cinereae) on the gravel deposits of the low raised beaches. Rush pastures and sedge mires occupy wet depressions.

With the exception of some fine-textured estuarine alluvial deposits, for example, around Beauly, where a moderate range of crops is possible, most of the cultivated areas are capable of only a narrow range of crops. Where the soil texture is coarse and windblow is a serious problem, land use is restricted to improved grassland and rough grazings. Through the use of specialized planting techniques and careful selection of species, afforestation at Findhorn, Maviston and Lossiemouth has been highly successful.

Map unit 420   Occupying 69 square kilometres (60 per cent of the association), this unit occurs along the Moray Firth, for example, at Spey Bay, Lossiemouth, Findhorn, Maviston and Ardersier, and around Easter Ross. In general, the topography is flat or gently undulating and humus-iron podzols are widespread. Gleys are of only local significance. Within areas of windblown sand, high steep-sided dunes occur and noncalcareous regosols form the dominant soil type. Soil texture varies according to the type of the raised beach deposit. Estuarine alluvium often has a fine sandy loam or silty loam texture in contrast with the sands of normal beach deposits and with the shingle deposits which are characteristic of recent storm beaches.

The estuarine alluvial deposits of Easter Ross are of agricultural significance; for example, land capable of a moderate range of crops occurs around the Beauly Firth. Elsewhere along the Moray Firth the land use is restricted to improved grassland or rough grazings.

Map unit 421   This unit occupies 50 square kilometres of extensive depressions and deltas within the same geographical areas. Noncalcareous gleys are dominant with lesser areas of calcareous and peaty gleys, their wetness resulting from a high ground-water table.

In Easter Ross, for example, around Nigg Bay, these soils have been extensively drained and support arable crops, but elsewhere along the Moray Firth most of the unit is restricted to rough grazing.

THE NOCHTY ASSOCIATION
(Map unit 422)

The soils of map unit 422 belong to the Nochty Association and are formed from fluvioglacial gravels, sands and silts derived from basic rocks. The unit also includes mineral alluvial soils derived in part from such basic rocks.

It is a minor unit, occurring only on terraces and mounds in Strathdon valley, and covers 10 square kilometres, less than 0.1 per cent of the land area. It is found between altitudes of 250 and 380 metres. Slopes are usually less than 15 degrees. The land is non-rocky and non-bouldery.

The average annual rainfall is 1000 millimetres within the climate region of cool, wet foothills and uplands. The plant communities are arable and permanent pastures (Lolio-Cynosuretum), acid bent–fescue grassland (part of Achilleo-Festucetum tenuifoliae) and dry boreal heather moor (part of Vaccinio-Ericetum cinereae).

Freely drained brown forest soils and mineral alluvial soils are dominant. Many of the mineral grains are dark in colour because of their derivation from basic rocks. This feature contrasts with the paler colour of most mineral grains in soils of the Corby and Boyndie Associations which are derived mainly from acid rocks. Development of brown forest soils, rather than the predominant podzols of the Corby and Boyndie Associations, is a consequence of the higher base status of the parent materials in similar situations.

The land is used for arable agriculture, rough grazing and coniferous plantations, the main limitation being climate. There is considerable potential for reclamation to long ley grassland.

THE NORTH MORMOND/ORTON ASSOCIATIONS
(Map units 423–426)

These associations occupy 233 square kilometres (0.9 per cent of the land area) around the Moray Firth from Evanton to Crimond, near Rattray Head. The parent materials are drifts derived from Old Red Sandstone sediments, mainly conglomerates and sandstones, and acid metamorphic rocks. West of the River Spey, the latter are Moine schists and granulites whereas to the east they are mainly Dalradian quartzites and schists. Locally, granite is also found in the drift. The acid metamorphic stones are often subrounded and they may have been derived originally from the conglomerate.

The North Mormond Association forms a belt 10 kilometres long on the north side of Mormond Hill. The parent material is a reddish brown sandy clay loam till which, below 50 metres altitude, has been partially water-sorted; the resultant texture is often a coarse loamy sand.

The Orton Association occurs between Keith and the River Nairn and the parent material is normally a reddish brown, compact sandy loam till.

West and north of Inverness the soils have not been assigned to either association though they have a close resemblance to those of the Orton Association. Though these soils show considerable variation over a short distance the parent material is usually a pale brown, stony, sandy loam or loamy sand. Locally, the parent material is a reddish brown sandy clay loam till, the finer texture either reflecting underlying sandstones and shales or indicating a bisequal deposit.

Landforms are mainly lowlands and foothills with gentle and strong slopes. In a large area west of Inverness, they also comprise hill slopes which are strong or steep and rise to a maximum altitude of 300 metres.

The average annual rainfall is between 750 and 1150 millimetres and the climate regions range from sheltered, warm moderately dry lowlands to cool wet foothills and uplands.

Humus-iron podzols are the dominant soils; brown forest soils also occur but are restricted to the Aberdeenshire lowlands. Gleys occur scattered throughout the association area.

Arable and rotational grassland are the main forms of land use though there are large areas devoted to forestry and rough grazings, especially west of the River Spey and north and west of Inverness.

Map unit 423   Brown forest soils dominate this unit which is restricted to a 10-kilometre belt north of Mormond Hill in Aberdeenshire; it covers 21 square kilometres (9 per cent of the associations).

The parent material is a reddish brown, sometimes a strong brown, sandy clay loam till generally more than 1 metre thick and containing quartzite, quartz-schist, andalusite-schist and granite stones. The derived brown forest soils are usually imperfectly drained but the soils developed on the water-sorted till with readily leached upper horizons are mainly transitional to humus-iron podzols. Minor areas of freely drained humus-iron podzols do occur locally. Poorly drained noncalcareous gleys occupy hollows and receiving sites.

Virtually all this unit is under arable agriculture or rotational grassland and is capable of producing a moderate range of crops; climate is the principal limiting factor.

Map unit 424   The unit consists of noncalcareous gleys with some peaty and humic gleys. It occupies gently sloping sites within undulating lowlands and foothills. Scattered throughout the association area, the unit covers 34 square kilometres (15 per cent of the associations).

Generally, the noncalcareous gleys predominate though peaty and humic gleys, sometimes accompanied by peat, are common west of the Great Glen.

Associated plant communities are arable and permanent pastures (Lolio-Cynosuretum) with rush pastures and sedge mires and bog heather moors (Narthecio-Ericetum tetralicis) on the uncultivated land.

Outwith Aberdeenshire, where the unit is capable of producing a moderate range of crops, the unit is only marginal for arable agriculture and is best suited to the production of a narrow range of crops based on long ley grassland. A substantial area, especially at the higher elevations, is devoted to rough grazing. Soil wetness and the peaty surface layers are the main limitations.

Map unit 425   This is the most extensive unit, occupying 173 square kilometres (74 per cent of the associations) around the lowland fringe from Evanton to Keith. The main area lies between Inverness and Marybank on the non-rocky foothill slopes which are mostly regular and strong, rising to 250 or 300 metres over a distance of 2 to 4 kilometres. Other areas consist of both non-rocky undulating lowlands and foothills.

Humus-iron podzols are the dominant soils with peaty podzols and gleys as minor components. Most of the humus-iron podzols are cultivated, especially on the lower slopes. Peaty podzols occur locally at the higher elevations where the gradient is often less. Gleys, mainly noncalcareous, are confined to wet channels and hollows. The parent material is most commonly a brown sandy loam, generally indurated and occasionally overlying a compact, reddish brown sandy loam till.

The plant communities include arable and permanent pastures (Lolio-Cynosuretum) with rush pastures and sedge mires on the restricted wet soils. Boreal and Atlantic heather moors (Vaccinio-Ericetum cinereae and Carici binervis-Ericetum cinereae) dominate the uncultivated higher ground.

Most of the low ground is devoted to arable agriculture and where many of the soils are imperfectly drained, for example, around Kirkhill and Marybank, it is capable of producing a moderate range of crops. On the lower and middle slopes of the foothills, where the gradient is limiting, the land is capable of producing only a narrow range of crops based on grassland. These are areas of mainly small farms and crofts. Most of the upper slopes are afforested or used for rough grazing. This land is only moderately suited to grassland reclamation because of the combined limitations of soil pattern and stoniness, and locally, a peaty surface layer.

Map unit 426   With a single area of 5 square kilometres (2 per cent of the associations) the unit occurs on the hills north of Rothes.

Peaty podzols cover the gently sloping hill tops which rise to 260 metres. Peaty gleys occupy minor receiving sites and peat infills streamhead channels. On the lower, steeper slopes, there are imperfectly drained podzols and noncalcareous gleys.

Almost all the unit is planted to conifers. The minor, unplanted areas support Atlantic and boreal heather moors (Carici binervis-Ericetum cinereae and Vaccinio-Ericetum cinereae) with wet pastures and sedge mire in the wet sites. The land is moderately suited to grassland reclamation.

THE ORDLEY/CUMINESTOWN ASSOCIATIONS
(Map units 427 and 428)

The soils of both the Ordley and Cuminestown Associations are developed on drifts derived mainly from Middle Old Red Sandstone sediments; that of the Ordley Association contains argillaceous schist rocks derived from the basal conglomerate beds of the Old Red Sandstone formation. Together the associations cover 166 square kilometres or 0.7 per cent of the land area, the parent material of the Ordley Association being the more extensive. Soils of the group, mainly the Ordley Association, occur over a belt of country from 3 to 9 kilometres wide stretching from Gardenstown on the north coast, southwards by Turriff, to Fyvie. Soils of the Cuminestown Association occupy a smaller area which follows the Bogie valley from the vicinity of Rhynie and stretches southwards to slightly beyond Kildrummy. Another minor area extends over Cotton Hill, 5 kilometres south of Cullen.

Two map units have been distinguished in these associations, which occur in climate regions ranging from fairly warm, moderately dry lowlands and foothills with an average annual rainfall of 850 to 1000 millimetres to cool, wet foothills and uplands with 1000 millimetres rainfall. Altitude ranges from 90 to 195 metres.

Map unit 427   With a total area of 12 square kilometres (5 per cent of the associations), this unit consists of noncalcareous and peaty gleys. It occurs in depressions and on gentle slopes of undulating lowlands in several scattered areas to the east and north-east of Turriff. The soils generally have a dark reddish grey to very dark brown loam topsoil with a coarse subangular blocky structure and a moderate to high content of stones, mainly red sandstones with some schists. Both B and C horizons, which are reddish brown to reddish grey, have sandy clay loam textures and there is generally an increase in the content of sandstones. The structure in the B horizon is strong prismatic, but the C horizon is massive. Much of the area is under arable cultivation with wetness, and sometimes texture, being the main limitations. It produces a moderate range of crops. An area immediately north of Newbyth includes patches of very poorly drained peaty gleys and is limited to long ley grassland. Uncultivated areas are in permanent pasture or support rush pastures and sedge mires.

Map unit 428   This unit covers 154 square kilometres (95 per cent of the associations) and comprises brown forest soils and humus-iron podzols with some noncalcareous gleys. It occurs widely throughout the districts covered by the association group and is found on undulating lowlands with gentle and strong slopes. The dominant soils are freely and imperfectly drained and have a reddish brown loam topsoil with medium subangular blocky or crumb structure; those of the Ordley Association have a high content of argillaceous schist stones. The B horizon is a reddish yellow or reddish brown loam with subangular blocky structure, whereas the underlying C horizon is a reddish brown sandy loam or loam which is normally compact and massive. In the Ordley Association both B and C horizons contain many rounded chips of argillaceous schist whereas in the Cuminestown Association the sandstone content steadily increases with depth.

   The stone content of the soil is generally insufficient to interfere with cultivation and, in the area from Gardenstown southwards to Fyvie, much of the unit is cultivated and produces a moderate range of crops with high or moderate yields. In areas where the slopes are sufficiently steep to provide a limitation, the range and yields of crops are reduced. This situation occurs also where the climate is cooler, for example, in the Rhynie–Kildrummy district. Where slope is an added limitation in these cooler areas, the land use is restricted further to a narrower range of crops based on long ley grassland.

THE PETERHEAD ASSOCIATION

(Map units 429 and 430)

The Peterhead Association is developed on drifts, mainly of clay loam or clay texture, derived from Old Red Sandstone sediments with igneous and metamorphic rocks and some conglomerate cobbles. It covers a total area of 138 square kilometres (0.6 per cent of the land area) and occurs, together with enclaves of red lacustrine clay of the Tipperty Association, along the coastal plain from Crimond, north of Peterhead, to Foveran, north of Aberdeen. Two map units have been distinguished throughout a climate region of fairly warm, moderately dry lowlands and foothills with an average annual rainfall of 760 millimetres.

Map unit 429   This map unit covers 71 square kilometres (50 per cent of the association) and consists of brown forest soils with gleying and some gleys. It is extensive from Peterhead southwards to Foveran and occurs in undulating lowlands with gentle and strong slopes in an altitude range of 30 metres to 45 metres. The soil profile has a reddish brown clay loam topsoil with medium or coarse crumb structure overlying a reddish brown clay loam or clay subsoil with prismatic structure which changes to massive below approximately 80 centimetres. Although the clayey texture of the soil has led to poor drainage, improvements have been effected by tile-draining in most areas and much of the unit provides good arable land capable of producing a moderate range of crops.

Map unit 430   Covering 67 square kilometres (50 per cent of the association), it comprises noncalcareous gleys with some peaty gleys and brown forest soils with gleying. It occurs in various parts of the association area where the topography is level or gently undulating. The unit closely resembles *map unit 429* although the textures of the subsoil horizons tend to be slightly finer in map unit 430 and the structure faces are strongly gleyed. Most of the area is cultivated and produces a moderate range of crops. Wetness is the main limitation, though high yields of grass, oats and barley are possible. Somewhat higher yields, over a wider range of crops, are found in a few areas, generally bordering or within areas of *map unit 429*.

THE ROWANHILL/GIFFNOCK/WINTON ASSOCIATIONS
(Map units 444–448)

The three associations included in this group are all developed on materials derived from Carboniferous sediments. The parent material of Rowanhill Association is formed mainly from sandstones and shales, and that of the Giffnock Association from sandstones with some shales, coals and limestones. The Winton Association, distinguished by its reddish brown colour, was originally mapped in East Lothian, but the soils are now included as series of the Rowanhill Association. Covering a land area of 600 square kilometres (2.4 per cent of the land area), the group occurs extensively throughout the lowlands of Fife and stretches westwards to Alloa on the Forth Lowlands. It occurs within climate regions ranging from warm, moderately dry lowlands with an average annual rainfall of 650 millimetres to fairly warm, moderately dry lowlands and foothills with 1000 millimetres rainfall. Five map units have been distinguished in Eastern Scotland (Sheet 5).

Map unit 444   This unit is developed on water-sorted drift and consists of imperfectly drained brown forest soils with gleying and some freely drained brown forest soils. It is closely related to, and found in conjunction with, *map unit 445*, generally occurring on undulating lowlands with gentle and strong slopes. The soils of map unit 444 are coarse textured normally to a depth of more than 0.6 metres and in consequence are easily worked, with the underlying fine-textured drift ensuring an adequate supply of moisture in time of drought. They cover 92 square kilometres (15 per cent of the associations).

Most of the unit is devoted to arable agriculture and produces a moderate range of crops with high yields. Some areas are capable of a wider range though others are restricted to permanent pasture.

Map unit 445   This is the largest unit and occupies 444 square kilometres (74 per cent of the associations). It comprises brown forest soils with gleying and occurs on undulating lowlands and valley sides with gentle and strong slopes, together with some noncalcareous and humic gleys in depressions. The brown forest soil profile normally has a dark grey-brown loam to silty clay loam topsoil on a grey or light brownish grey sandy clay loam with frequent stones, mainly shales and sandstones. This overlies a dark grey-brown or dark grey clay loam drift containing frequent weathered sandstones and shales. Although texture is the main physical limitation affecting land use, the greater part of the unit produces a moderate range of crops. In certain areas, particularly in the coastal lowlands of East Fife where the climate is more favourable, higher yields over a wider range of crops are obtained. Further to the west, however, additional limitations, imposed mainly by gradient or pattern, progressively restrict land use to a narrow range of crops based on long ley grassland. A few small areas supporting rush pastures and sedge mires are suitable for grassland reclamation.

Map unit 446 The unit is not extensive and is found mainly in West Fife covering 49 square kilometres (8 per cent of the associations). It consists of noncalcareous gleys together with some brown forest soils with gleying and peaty gleys. These soils occur generally in depressions and hollows throughout areas of undulating lowlands with gentle and strong slopes. Much of the unit, which has wetness and textural limitations, is under arable cultivation and produces a moderate range of crops with average yields or is restricted to a narrow range based on long ley grassland. Several areas, particularly on peaty gleys or on reclaimed open-cast coal sites, may support rush pastures and sedge mires or broadleaved woodland, mainly elmwood (Querco-Ulmetum glabrae) and southern oakwood (Galio-saxatilis-Quercetum), but are moderately or marginally suited to grassland reclamation.

Map unit 447 The smallest of the five units, it occurs on mounds on valley sides mainly in drift-free areas of West Fife. The unit totals only 5 square kilometres (less than 1 per cent of the associations) and consists of brown forest soils developed largely on weathered sandstone or, occasionally, on shallow drift.

Map unit 448 This unit is found also only in West Fife and is developed on coarse-textured shallow drift or on residual sandstone. The soils are humus-iron podzols, both freely and imperfectly drained, with some brown forest soils and gleys. It is a minor unit and covers only 10 square kilometres (2 per cent of the associations). In most areas, this unit and *map unit 447* are suitable only for rough pasture or broadleaved woodland, mainly elmwood (Querco-Ulmetum glabrae) and southern oakwood (Galio saxatilis-Quercetum); in Fife, where the soils are slightly deeper, there are areas which produce a narrow range of crops based on long ley pasture.

THE SABHAIL/MOUNT EAGLE ASSOCIATIONS

(Map units 454–456 and 457)

Soils of the Sabhail/Mount Eagle Associations are developed on drifts derived mainly from sandstones of Lower and Middle Old Red Sandstone age. They occur in Easter Ross, the Black Isle and on Drummossie Muir and cover 129 square kilometres (0.5 per cent of the land area).

The parent sandstone rocks range in colour from grey, brown and yellow to red. Flagstones with occasional red, black and green shales underlie part of Drummossie Muir in addition to the sandstones. Shales also occur locally in other areas, for example, south of Edderton. The parent material of the Mount Eagle

Association is usually a reddish brown and coarse-textured, stony, sandy loam. Shattered sandstone is found normally less than 50 centimetres from the surface. In the Sabhail Association, a compact, brown sandy loam till is the parent material.

Freely drained humus-iron podzols are dominant in the Mount Eagle Association whereas peaty podzols are dominant in the Sabhail Association. On gentle slopes, the peaty podzols often have a well-developed iron pan with the overlying horizons being poorly drained. Peaty gleys, noncalcareous gleys and peat are minor soils, normally confined to shallow depressions. Rankers are associated with steep, rocky slopes above Loch Ness.

Landforms include undulating lowlands, foothills and hills and valley sides. Slopes in the lowlands and foothills are gentle and strong and non- or slightly rocky. On the hill and valley sides, the slopes vary from gentle to steep and from non-rocky to moderately rocky.

Climate regions, though including moderately dry lowlands, are mainly fairly warm moderately dry lowlands and foothills and fairly warm wet lowlands and foothills.

Most of the lowlands are cultivated whereas much of the higher ground is devoted to forestry and rough grazing; there is a tendency towards further afforestation.

Map unit 454   This is the most extensive unit and covers 105 square kilometres (85 per cent of the associations). Freely drained humus-iron podzols, both cultivated and uncultivated, occur on non- or slightly rocky, undulating lowlands and foothills; slopes are gentle and strong. Gleys, mostly noncalcareous and peaty, are restricted usually to hollows. Peaty podzols exist on gentle slopes above 200 metres.

The largest area of the unit is on Drummossie Muir where it occupies a belt 3 kilometres wide between altitudes of 200 and 300 metres. Although the overall slope is gentle, the unit west of the A9 road consists of a complex of low parallel ridges; they are aligned south-west to north-east. The ridges are formed in till, moraine or shattered rock. Humus-iron podzols and some peaty podzols, often imperfectly drained, occupy the ridges, with peat and peaty gleys in the hollows.

Most of the area is cultivated up to 150 metres. Above this limit, land use is mainly rough grazing and forestry. The dominant plant communities are arable and permanent pastures (Lolio-Cynosuretum) with rush pastures and sedge mires in the associated hollows. Dry and moist Atlantic heather moors (parts of Carici binervis-Ericetum cinereae) dominate the humus-iron and peaty podzols.

Agricultural capability ranges from small areas capable of a wide range of crops, for example, near Evanton, to land which is only moderately suited to grassland reclamation. Climate, stoniness and induration are the main limiting factors.

Map unit 455   This map unit was included on the map legend by error. It consists of peaty podzols with some gleys and humus-iron podzols and occurs as a single area (51 square kilometres) in Morangie Forest, west of Tain, in Northern Scotland (Sheet 3).

Map unit 456   The unit covers 6 square kilometres and consists of non-rocky hills with gentle and occasional strong slopes. Only one area occurs in Eastern Scotland (Sheet 5), situated north of Loch Duntelchaig at an altitude between 200 and 310 metres.

Peaty podzols and peat are the dominant soils; minor components are peaty gleys. Dry and moist Atlantic heather moors (parts of Carici binervis-Ericetum cinereae) are the principal plant communities on the podzols. Lowland and upland blanket bogs (parts of Erico-Sphagnetum papillosi) and bog heather moors (Narthecio-Ericetum tetralicis) characterize respectively the peat and peaty gleys.

Land use is devoted mainly to rough grazing with increasing afforestation although some grassland reclamation has been undertaken. Much of the land is suited to grassland improvement.

Map unit 457   This unit occurs as a single elongated plateau covering 18 square kilometres (15 per cent of the association) south-west of Drumnadrochit. The plateau rises gently from 200 metres in the north-east to 350 metres in the south-west and is underlain by a red sandstone, probably of Lower Old Red Sandstone age. Slopes are generally gentle or strong and the unit is non- or slightly rocky. Low, rock-controlled ridges, aligned south-west to north-east, occur in several localities.

At the southern end, peaty podzols dominate the ridges with peaty gleys in hollows and on gentle slopes. Towards the north, there are cultivated patches of these soils. Subsoil textures are sandy loam and loamy sand and the soils tend to be shallow; induration occasionally induces surface-water gleying. On the very steep plateau slopes leading to Loch Ness, the soils are developed on colluvium; they include humus-iron podzols, brown forest soils and rankers. These slopes are moderately rocky.

On the podzols of the plateau, the main plant communities are dry and moist Atlantic heather moors (parts of Carici binervis-Ericetum cinereae); in the hollows are bog heather moors (Narthecio-Ericetum tetralicis). The steep colluvial slopes are characterized by bent–fescue grassland with bracken (part of Achilleo-Festucetum tenuifoliae) and eastern highland oak and birchwood (Trientali-Betuletum pendulae).

Land use in the southern section of the plateau is restricted mainly to rough grazings; these grazings are of low value. The arable areas in the northern sector are capable of producing a narrow range of crops; other areas in the north are suited to grassland reclamation. On the steep slopes, land use is divided between rough grazing and forestry with the grazings being of moderate and high values.

THE SKELMUIR ASSOCIATION

(Map units 462 and 463)

Soils of the Skelmuir Association are developed on a stony, sandy loam to clay loam drift, the stones consisting of rounded pebbles of quartzite and Cretaceous flint varying in size from 1 to 15 centimetres. The drift bears no relation to the underlying solid rock and its origin is uncertain. The association covers an area of 20 square kilometres or less than 0.1 per cent of the land area and occurs at altitudes between 60 and 135 metres to the south-west of Peterhead in the vicinity of the Moss of Cruden. It is probable that blanket peat at one time covered the greater part of the association area.

Two map units have been distinguished in the association, which lies in a climatic region of fairly warm, moderately dry lowlands and foothills with an average annual rainfall of 760 millimetres.

Map unit 462 This map unit covers 11 square kilometres (55 per cent of the association); it consists of noncalcareous gleys with some peaty gleys and occurs on level ground and on gentle concave slopes in undulating lowlands. The dominant soil is developed on a loam or clay loam drift and is poorly drained. The surface horizon is a black, stony loam with a high percentage of organic matter. The nutrient status of the soil tends to be low and the limitations imposed by wetness and by excessive stoniness render the land suitable for only average production of a moderate range of crops.

Map unit 463 The unit covers 9 square kilometres (45 per cent of the association) and consists of peaty podzols with minor noncalcareous gleys. It occurs on hills and valley sides and is normally found on strong and steep, convex slopes. The parent material is a sandy loam drift which frequently incorporates material from the underlying local rock. The uncultivated soil has a surface horizon of black greasy humus, a grey sandy loam E horizon and a thin iron pan, 2 millimetres thick, overlying sandy loam drift. Parts of these horizons may still be seen in the cultivated soil although in a few places all have been incorporated in the surface horizon which is normally black or very dark grey. The profile is poorly drained above the iron pan and freely or imperfectly drained below. Stoniness and, to a lesser extent, wetness are the main limitations to arable agriculture which is restricted to a moderate range of crops with average yields.

THE SORN/BIEL/HUMBIE ASSOCIATIONS
(Map units 466 and 467)

The soils of the Sorn, Biel and Humbie Associations are developed on drifts derived from a mixture of Lower Carboniferous, mainly calciferous sandstones and cementstones, and Upper Old Red Sandstone sediments, mainly sandstones and cornstones, and basaltic or andesitic lavas. Many of these rocks are reddish and impart that colour to the clayey till which is the main form of drift. Only soils of the Sorn Association are located in Eastern Scotland (Sheet 5).

The Sorn Association is not extensive and occupies only 5 square kilometres (less than 0.1 per cent of the land area). It occurs on the southern margin of the Forth valley between Kippen and Gargunnock at the junction of the till plain with the adjacent lava hills. Two units have been separated; map unit 466 accounts for 20 per cent of the association and map unit 467 for the remaining 80 per cent.

Average annual rainfall varies from 1200 to 1250 millimetres in the fairly warm, wet lowlands and foothills of the Forth valley margin. The dominant soils are noncalcareous gleys, with some brown forest soils with gleying on the better-drained sites.

Map unit 466 Occupying only 1 square kilometre, the unit consists almost entirely of imperfectly drained brown forest soils with gleying developed on compact, fairly massive, sandy clay loam and clay loam till moulded into drumlins. Minor areas of poorly drained noncalcareous and peaty gleys occupy the wet hollows between drumlins. The topsoil is loamy with a moderate structure, whereas the subsoil is usually a sandy clay loam with a moderate structure and moderate permeability in spite of an increased stoniness. The underlying till is almost massive with poor permeability and is the main reason why the soils are used mainly for permanent pasture (Lolio-Cynosuretum)

140

although arable agriculture is possible. Inadequate drainage results in rush-infested pasture. Where drainage is installed it should include permeable infill in the drain lines and a secondary treatment to improve the subsoil structure and permeability.

Map unit 467   This unit occupies 4 square kilometres (80 per cent of the association) and is dominated by poorly drained noncalcareous gleys with minor brown forest soils with gleying on the steeper slopes and humic and peaty gleys in the wettest hollows. All the soils are developed on a massive, poorly permeable, clay loam or clay till. The topsoils are loams or silty loams with generally a poor structure and a tendency to poach readily in the prevalent wet conditions, especially on gentler slopes with slow run-off of excess water. The principal land use is permanent pasture (Lolio-Cynosuretum), although arable cropping is possible. Rush-infested swards are common where drainage is inadequate and any scheme for drainage improvement must include permeable infill in the drain-lines and a secondary treatment to improve structure and permeability in the subsoil. There are minor areas of broadleaved woodland, usually southern oakwood (Galio saxatilis-Quercetum).

THE SOURHOPE ASSOCIATION

(Map units 472–476 and 479)

The Sourhope Association is developed on drifts derived from Old Red Sandstone intermediate lavas with some agglomerates and tuffs. The parent material varies from a brown or pale reddish brown, residual, stony sandy loam on the high ground to a brown or reddish brown loam or sandy clay loam on lower ground and in hollows. Covering a total area of 545 square kilometres (2.2 per cent of the land area), the association occurs widely on the Ochil Hills where it extends over an altitude range from 700 metres on Ben Cleuch, some 5 kilometres north of Alloa, to 90 metres above Newport on the River Tay. North of the river, it spreads over the south-west portion of the Sidlaw Hills between Kinfauns (220 metres) and Abernyte (115 metres) north of Kinnaird.

Brown forest soils are widespread on lowlands and on lower and middle hill slopes; on some steeper slopes, rock occurs near the surface. Noncalcareous and peaty gleys are found in hollows and on gentle slopes. Humus-iron podzols occur on upper slopes and peaty podzols, frequently with an iron pan, are found above 300 metres. Climate regions range from warm, moderately dry lowlands with an average annual rainfall of 760 millimetres in the east to cold, wet uplands with 1800 millimetres rainfall in the west.

Map unit 472   Of the six units mapped in Eastern Scotland (Sheet 5), map unit 472 is the most extensive and covers 405 square kilometres (74 per cent of the association). It occurs on undulating lowlands bordering Strathallan and lower Strathmore, on mainly strong slopes of the Ochil Hills (Plate 14) and on the south-west portion of the Sidlaw Hills. The unit comprises freely drained brown forest soils, some imperfectly drained brown forest soils with gleying and gleys. The main parent material is a shallow, stony drift of gritty loam to sandy clay loam texture except on some upper slopes where the drift is absent or very thin and the soil is developed on decomposed lava rock.

Much of the lowlands and lower hill slopes is in arable cultivation and produces a moderate range of crops. In some areas, due to stoniness and topography,

Plate 14. *The Eastern Ochils south of Bridge of Earn. In the foreground, intensive arable agriculture is dominant on brown forest soils of the Old Red Sandstone drifts, map units 337 (Kippen Association, Classes 2 and 3) and 414 (Mountboy Association, Classes 2 and 3). On the slopes and summits of the lava hills, moderate grazings are developed on the brown forest soils of map units 472 and 479 (Sourhope Association, Classes 4 and 5).* Aerofilms.

cultivation is restricted to the production of a narrow range of crops based on long ley grassland. Upper slopes support acid bent–fescue grassland (part of Achilleo-Festucetum tenuifoliae) and are well suited to grassland reclamation, though small areas are capable of only rough grazing with high grazing values. Minor areas of gleys support rush pastures and sedge mires.

Map unit 473   This unit is not extensive and covers only four small areas in the Ochil Hills totalling 4 square kilometres (less than 1 per cent of the association). The main constituents are noncalcareous gleys, some brown forest soils with gleying and peaty gleys. In two of the areas, pockets of peat are also found. Occurring on undulating foothills with gentle slopes, at altitudes around 420 metres, the soils support mainly rush pastures and sedge mires with some flying bent grassland (part of Junco squarrosi-Festucetum tenuifoliae). Owing to the limitations of wetness and altitude, the land is only moderately suited for grassland reclamation or provides rough grazings which have moderate grazing values.

Map unit 474   This unit covers a small area of 12 square kilometres (2 per cent of the association) at the western end of the Ochil Hills above Bridge of Allan. Occurring on hill ground with strong and steep slopes, the unit consists mainly of shallow, brown forest soils with rock near the surface and some brown forest soils with gleying. Although parts are under arable cultivation, much of the area supports herb-rich bent–fescue grassland (part of Achilleo-Festucetum tenuifoliae) and broadleaved woodland mainly elmwood (Querco-Ulmetum glabrae) and southern oakwood (Galio saxatilis-Quercetum). Some of the area is moderately suited for the establishment of grass. Shallowness of soil, or rockiness, or both, together with steepness, limit the remaining areas to rough grazings.

Map unit 475   Humus-iron podzols with some brown forest soils and peaty gleys make up map unit 475 which occurs in a few small areas of the Ochil Hills at altitudes above 300 metres. It covers 8 square kilometres (2 per cent of the association). Slopes are gentle and strong and the soils support dry Atlantic heather moor (part of Carici binervis-Ericetum cinereae), acid bent–fescue grassland (part of Achilleo-Festucetem tenuifoliae) and common white bent grassland (part of Junco squarrosi-Festucetum tenuifoliae). Apart from a small area which occurs on one lower hillslope and is capable of producing a narrow range of crops based on long ley grassland, the unit is suited only for reclamation to grassland and for rough grazings.

Map unit 476   This is the second most extensive map unit of the association, covering 72 square kilometres (13 per cent of the association). It is found on strong and steep slopes of the Ochil Hills at altitudes above 300 metres and with an average annual rainfall of over 1000 millimetres. The main constituents are peaty podzols which have a peaty surface horizon of variable thickness and a thin iron pan; imperfectly or poorly drained in the surface layers, the soils are freely drained below the pan. Minor components of the unit are humus-iron podzols and peat. The vegetation is dry Atlantic heather moor (part of Carici binervis-Ericetum cinereae) and common white bent grassland (part of Junco squarrosi-Festucetum tenuifoliae) on the drier areas with blanket bogs (Erico-Sphagnetum papillosi) where the drainage is poor. Altitude and, in some areas, topography or wetness, or both, restrict land use to rough grazings which have a moderate grazing value.

Map unit 479   Brown forest soils, which occur generally on moderately rocky hill sides with very steep slopes, are the main constituents of map unit 479. It also includes some brown rankers, humus-iron podzols and humic gleys. The unit is most extensive along the scarp of the Ochil Hills, stretching above the Devon valley from Bridge of Allan to Alloa, and covering 44 square kilometres (8 per cent of the association). Other smaller areas are found on the steep sides of several glens through the Ochil Hills. Because of rock and slope limitations, the soils of the unit are uncultivated and are in permanent pastures (Lolio-Cynosuretum) or support acid bent–fescue grassland (part of Achilleo-Festucetum tenuifoliae) or locally, herb-rich or dry Atlantic heather moors (parts of Carici binervis-Ericetum cinereae). Some areas are moderately or marginally suited to grassland reclamation, but other areas are restricted to rough grazings which have high, moderate, or low grazing value. Occasional areas of broadleaved woodlands consist mainly of elmwood (Querco-Ulmetum glabrae) and southern oakwood (Galio saxatilis-Quercetum).

143

THE STIRLING/DUFFUS/POW/CARBROOK ASSOCIATIONS

(Map units 487 and 488)

The four associations comprising this group are all developed on estuarine and lacustrine raised beach silts and clays. Collectively, they cover 240 square kilometres (1 per cent of the land area).

Stirling Association, by far the most extensive, covers soils formed on grey estuarine deposits of silty clay known as the 'Carse clays'. They are the Carses of Gowrie (Plate 15) and Earn bordering the River Tay and the Carse of Stirling bordering the River Forth. The Carbrook Association is much less extensive and is developed on reddish brown, high raised beach silts and clays occurring approximately 25 metres upslope from the carse deposits and also as an isolated area in Stratheden. Confined to a small area on the western side of the Montrose Basin is the Pow Association. The estuarine alluvium which forms the parent material is generally of a coarser texture than that of the Stirling Association, being silty loam or silty clay loam. The soils of the Duffus Association, which is the least extensive of the four, cover approximately 5 square kilometres west of Loch Spynie in Morayshire. Dark grey-brown, lacustrine silty clay forms the parent material; it is generally calcareous.

The soils are poorly drained noncalcareous and calcareous gleys and brown forest soils with gleying. Minor areas of peaty gleys and peat also occur.

Landforms are exclusively level or very gently sloping raised beaches and terraces, mainly below 10 metres; those of the Carbrook Association occur between 15 and 35 metres.

The average annual rainfall is 650 to 700 millimetres in Morayshire and Stratheden, 700 to 750 millimetres in the Carses of Gowrie and Earn and from 900 to 1100 millimetres around Stirling.

Almost all the land is cultivated, with the Loch Spynie area and parts of the Carse of Gowrie capable of producing a wide range of crops with high yields. The other areas are capable of producing only a moderate range of crops though the Carse of Stirling, with its heavier rainfall and higher clay content, is renowned for grass seed production, especially Timothy.

Map unit 487   The unit covers only 24 square kilometres (10 per cent of the association). It normally occurs adjacent to the clay soils of *map unit 488* but at slightly higher levels.

The soils are developed on estuarine deposits which contain slightly higher contents of both fine sand and silt than *map unit 488*. They comprise brown forest soils with gleying and some noncalcareous and calcareous gleys.

The landform is entirely that of raised beach terraces at altitudes between 15 and 35 metres; slopes are level and gentle.

Principal plant communities are arable and permanent pasture (Lolio-Cynosuretum), with rush and sedge mires associated with the gleys. Most of the land is cultivated and capable of producing a moderate range of crops; high yields of grass are common in the western areas.

Map unit 488   Covering 216 square kilometres, the unit occurs on raised beach terraces mainly below 10 metres. The textures of the parent materials, though mainly fine, show considerable variation. For example, in the Carses of Gowrie and Earn the clay contents vary from 40 to 60 per cent; in the Carse of Stirling the clay content is even higher, as much as 70 per cent. In the Duffus area,

Morayshire, clay contents vary from 35 to 50 per cent with silt figures of the same order or slightly lower. Around the Montrose Basin, clay contents are generally below 30 per cent with the silt figures being invariably higher. In a few minor areas, poorly drained soils developed on raised beach fine sands and silts have been included in the unit.

Plate 15. *The Tay valley, east of Perth. The soils on the terraces either side of the river are noncalcareous gleys, map unit 488 (Stirling Association, Class 3) and brown forest soils, map unit 89 (Carpow Association, Class 2). Soils of the hills and slopes beyond are developed on lavas and are dominated by brown forest soils of map unit 472 (Sourhope Association, Classes 3, 4 and 5).* Photograph ADS Macpherson.

The soils are mainly noncalcareous and calcareous gleys with some peaty gleys and peat. Because of the silty clay or clay texture, these mineral soils are massive and sticky when wet but develop a prismatic structure on drying out. The method and timing of cultivation are all-important for the creation of a good tilth.

Level and gently sloping, raised beach terraces comprise the only landform.

The land is mostly cultivated, and arable and permanent pastures (Lolio-Cynosuretum) are the main plant communities. In the Carses of Gowrie and Earn, and in Stratheden, the land is capable of producing a moderate range of crops; yields of cereals are high. Within the higher rainfall of the Carse of Stirling, the land, though capable of producing a moderate range of crops with moderate yields, is mainly devoted to the production of grass; yields are high. In the Loch Spynie area, the land is capable of growing a wide range of crops but is associated mainly with the production of high yields of wheat and barley.

145

THE STONEHAVEN ASSOCIATION
(Map unit 490–496)

Soils of the Stonehaven Association are developed on drifts derived from Lower Old Red Sandstone conglomerates and lavas. In Eastern Scotland (Sheet 5), the association covers a total of 253 square kilometres (1.0 per cent of the land area). The largest area is in Kincardineshire where it occupies the rolling to hilly country between Stonehaven and the River North Esk, stretching for 29 kilometres with an altitudinal range of 30 metres to 240 metres. Further south, it occurs at intervals along the line of the Highland Boundary Fault in a relatively narrow zone, 1 to 2 kilometres broad, stretching from Stroness Hill north-east of Crieff to Lurgan Hill south-east of Comrie, and extending from Loch Venachar to Loch Lomond. The altitude ranges from 30 metres on the shores of Loch Lomond to 460 metres on the Perthshire hills.

Seven map units have been distinguished, occurring in climate regions ranging from warm, moderately dry lowlands with an average annual rainfall of 700 millimetres in the east through fairly warm, moderately dry lowlands and foothills with 760 millimetres rainfall to warm, wet lowlands with 1300 millimetres and cool, wet foothills and uplands with 2000 millimetres in the south-west.

Map unit 490   This is the most extensive of the seven map units. It covers 180 square kilometres (71 per cent of the association), mainly in Kincardineshire and comprises imperfectly drained and freely drained brown forest soils with some humus-iron podzols and noncalcareous gleys. The dominant component is an imperfectly drained brown forest soil with gleying which has a brown, sandy clay loam topsoil overlying a weak-red, sandy clay loam with coarse angular blocky structure on a reddish brown, clay loam drift with coarse blocky or massive structure. Conglomerate cobbles, generally up to 8 centimetres in diameter, are common throughout the profile, particularly in the surface horizon where they can hinder cultivation. The drift is calcareous below about 2 metres. Most of the unit is arable land; there are limited areas of permanent pasture (Lolio-Cynosuretum) and broadleaved woodland.

Much of the unit, mainly where it occurs in the Mearns area of Kincardineshire, is cultivated and produces a moderate range of crops. Yields are often high. In some localities where topography or stoniness or both are added limitations, the land use is restricted to a narrower range of crops based primarily on grassland. In some hill areas, particularly in the south-west, the land is only moderately, or marginally, suited to grassland reclamation because of altitude and slope. Occasionally, the restrictions are so severe that land use is limited to rough grazing. Such grazings are of high grazing value. Scattered, minor areas of broadleaved woodland comprise mainly elmwood (Querco-Ulmetum glabrae), eastern highland oak and birchwood (Trientali-Betuletum pendulae) and Atlantic oakwood and western highland birchwood (Blechno-Quercetum).

Map unit 491   Covering 10 square kilometres (4 per cent of the association), the unit consists of noncalcareous and peaty gleys with some humic gleys. It occurs among the Garvock Hills south of Stonehaven, mainly in depressions and on gentle, occasionally strong, slopes. Drainage is poor, in some localities very poor, and the unit has a greyish brown or dark greyish brown topsoil of sandy clay loam texture. As in *map unit 490*, conglomerate cobbles are common. Some profiles

146

show signs of water-sorting with topsoils of sandy loam, up to 30 centimetres thick, overlying sandy clay loam.

Much of the unit is arable or in permanent pasture (Lolio-Cynosuretum); some areas, where the drainage is very poor, support rush pastures and sedge mires.

In other areas, map unit 491 is restricted mainly by wetness to long ley grassland with only a limited potential for other crops: in some places where drainage has been improved, the land is under arable production and produces a moderate range of crops.

Map unit 492   This unit consists mainly of brown forest soils with some gleys and brown rankers and occasional rock outcrops. It occurs to the west and south-west of Crieff on undulating foothills with gentle and strong slopes, covering 26 square kilometres (10 per cent of the association). Much of the unit supports acid bent–fescue grassland (part of Achilleo-Festucetum tenuifoliae) with areas of broadleaved woodland, mainly elmwood (Querco-Ulmetum glabrae) and eastern highland oak and birchwood (Trientali-Betuletum pendulae); rush pastures and sedge mires are found on the gleys.

Much of the unit is limited, mainly by soil depth and by gradient, to rough grazings which are of high and moderate grazing value. To the south of Crieff, where the soil is deeper however, the land produces a narrow range of crops based primarily on grassland.

Map unit 493   Comprising humus-iron podzols with some peaty podzols on shallow drift and gleys, it covers 29 square kilometres (12 per cent of the association). Occurring on undulating lowlands and foothills with gentle and strong slopes, the unit spreads over much of the Lurgan Hills, south-west of Crieff, and the Menteith Hills, south-west of Callander. Although the land is generally non-rocky, outcrops do occur occasionally. The unit is uncultivated and supports dry Atlantic heather moor (part of Carici binervis-Ericetum cinereae), acid bent–fescue grassland and common white bent grassland (parts of Achilleo-Festucetum tenuifoliae and Junco squarrosi-Festucetum tenuifoliae) respectively.

Map unit 493 is limited by topography and by shallowness of soil to rough grazings with only a small area marginally suited to grassland reclamation; the rough grazings are of moderate and low grazing value.

Map unit 494   This unit, which covers only 3 square kilometres (1 per cent of the association), has peaty podzols as its main component with humus-iron podzols and gleys as minor constituents. It occurs on hill and valley sides with strong and steep slopes and, apart from occasional outcrops, is non-rocky. Plant communities are dry and moist Atlantic heather moors (parts of Carici binervis-Ericetum cinereae), bog heather moors (Narthecio-Ericetum tetralicis) and heath rush–fescue grasslands (Junco squarrosi-Festucetum tenuifoliae). Map unit 494, limited mainly by gradients, is suitable only for use as rough grazings; grazing values are low.

Map unit 495   Also covering an area of 3 square kilometres (1 per cent of the association), this unit consists of brown forest soils, some humus-iron podzols, brown rankers and gleys. It occurs on hills and valley sides with strong and steep, rounded slopes and is moderately rocky with frequent rock outcrops near summits. The unit supports acid bent–fescue grassland (part of Achilleo-Festucetum tenuifoliae) and broadleaved woodlands, mainly Atlantic oakwood and western highland birchwood (Blechno-Quercetum), with rush pastures and sedge mires on the wetter areas. The unit is restricted mainly by steep slopes and rockiness to rough grazing which has a high grazing value.

Map unit 496   This unit, the smallest of the seven, though closely related to the previous unit, has humus-iron podzols as the major component with brown forest soils, rankers and gleys as minor constituents. It occurs in only two areas, on Gualann Hill, south-west of Aberfoyle, and on Conic Hill beside Loch Lomond, covering 2 square kilometres (less than 1 per cent of the association). The landform is similar to that of *map unit 495* and consists of hills and valley sides with strong and steep, rounded slopes. There are frequent rock outcrops near hill summits. Plant communities consist mainly of dry Atlantic heather moor (part of Carici binervis-Ericetum cinereae), with some acid bent–fescue grassland and common white bent grassland (parts of Achilleo-Festucetum tenuifoliae and Junco squarrosi-Festucetum tenuifoliae respectively). This unit, occurring at higher altitudes than *map unit 495*, is capable only of use as rough grazing; the grazing values are low, occasionally moderate.

THE STRICHEN ASSOCIATION
(Map units 497–509, 512–515)

Within Eastern Scotland (Sheet 5), the Strichen Association is the largest association and covers 3827 square kilometres (15.6 per cent of the land area). It spans a wide range of altitude from 100 to 900 metres and covers considerable variation of soil type, topography and vegetation. This accounts for the identification of seventeen soil map units.

The soils are developed on parent materials derived from schists of the Dalradian Assemblage. Reference to the published 1:63 360 soil maps shows the soil parent materials have been amalgamated under the general heading of drift or have been divided into three categories, namely till, shallow drift and deeply weathered rock. Further identifiable types include colluvium, moraine and mountain-top detritus. Till deposits are widespread in lowland and foothill sites; in the latter, these deposits form gentle, straight or concave slopes. The till varies from a yellowish brown, sandy loam to a brownish yellow, sandy clay loam. Evidence of water-sorting is generally absent. The stones are subangular and subrounded schists with occasional other rock types in localized areas. On hilltops and upper convex slopes, the soils are developed commonly on shallow drift overlying shattered rock. The characteristic textures are stony, sandy loam and loamy sand. Shattered rock is encountered often within 1 metre from the surface and frequently the local derivation of the parent material is indicated by the high content of angular stones. On mountain summits, intense frost action has caused physical disintegration of the rock *in situ* and the resultant mountain-top detritus has a very stony, sandy loam or loamy sand texture. Towards the base of the solum, the stones have distinct silty or fine sandy cappings which increase in thickness towards the underlying rock. Examples of deeply weathered rock are found most frequently in localized upland and mountain sites where the mantle of drift is thin or absent; textures are loam or sandy loam and there is a wide range of variegated mottles which result from the process of weathering. On steep slopes, colluvial deposits occur, characterized by a loam or sandy loam texture, a low stone content and the absence of induration. The material forming the morainic mounds has a sandy loam or loamy sand texture, a high content of subrounded or rounded stones, and is often indurated. The characteristic bedding

148

and stratification of the fluvioglacial deposits found, for example, in the Corby Association, is absent.

Average annual rainfall ranges from 850 millimetres in the fairly warm moderately dry lowlands of the north-east to 1200 millimetres in the cold wet foothills and to 1600 millimetres in the cold wet uplands of the North-East Grampian Highlands. In the South-West Grampian Highlands, the rainfall reaches a maximum of 3200 millimetres around upper Glen Lyon.

Stretching in a continuous belt from the Buchan Platform in the north-east to the Trossachs in the south-west, the association is extensive. Most of the commonly occurring soil subgroups are represented, though brown forest soils with free or imperfect drainage are restricted to the lowlands and some foothill sites. Of the soils with poor or very poor drainage, noncalcareous and peaty gleys are most common. Humus-iron podzols are predominant in the fairly warm and warm, moderately dry or wet lowlands and foothills, with peaty podzols predominant at higher elevations. Subalpine and occasional alpine soils are widespread above about 550 metres, the alpine soils usually occurring above 700 metres. On rugged landforms, rankers and lithosols are extensive.

A large proportion of the association is cultivated. On uncultivated land, plant communities are generally boreal heather moors (Vaccinio-Ericetum cinereae) or Atlantic heather moors (Carici binervis-Ericetum cinereae) on the free-draining podzols and bog heather moors (Narthecio-Ericetum tetralicis) on the podzols with poor drainage above an iron pan and on the peaty gleys. Birchwood (Trientali-Betuletum pendulae) and acid bent–fescue grassland (part of Achilleo-Festucetum tenuifoliae) are both found in sheltered valleys; coniferous plantations have a scattered distribution.

Nearly all the arable land is capable of producing only a moderate range of crops, the major limitation being climate which becomes moderately severe in the foothills where only a narrow range is possible. Above the limits of arable land, there are areas where grassland reclamation is possible. The upper hill slopes generally provide rough grazing of low value. Slopes below 550 metres usually have a considerable forestry potential.

Map unit 497   By virtue of its particular soil and topography relationship, this major unit has a scattered distribution, covering 430 square kilometres or 11 per cent of the association. In lowland sites, it is confined to footslopes and depressions; in the foothills, it comprises long, gentle, straight or concave slopes which may be often associated with a series of spring lines at the upper break of slope. Till is present in both lowland and foothill sites.

The soils are noncalcareous, peaty and humic gleys. Some humus-iron and peaty podzols with imperfectly drained B horizons are present on low mounds. Peat is generally absent. Within the lowlands, the gleys are often of the groundwater type, whereas in foothill sites, both ground-water and surface-water types are present, the latter being more frequent.

On lowland sites, the frequent occurrence of *Juncus*-infested areas and their marked contrast with drained areas, emphasize the need for adequate drainage systems. Following the long-term use of fertilizers, coupled with improved cereal varieties and suitable drainage, the natural infertility of these soils has been ameliorated and the production of a moderate range of crops is now possible. Where climatic conditions are more severe, however, similar land is capable of producing only a narrow range of crops. In some isolated valleys and remote areas, arable soils have been abandoned; these would respond successfully to surface reseeding following extensive drainage and fertilizer input.

Map unit 498   By far the most extensive unit in the association (938 square kilo-metres or 25 per cent of the association), it occupies large parts of the undulating lowlands such as the Buchan Platform, the 305-metre peneplain in Easter Ross, Morayshire and Nairnshire, and the lower slopes of upland areas such as the Cromdale Hills and the Ladder Hills; it occurs also in some foothills north of the Highland Boundary Fault, for example, in the South-West Grampian Highlands and in the Trossachs (Plate 16). In all these areas, till and shallow drift deposits are predominant, the latter on summits and upper convex slopes where there may be restricted rock outcrops.

Plate 16. *The slopes of the Ben Lawers group of mountains on Loch Tayside. There is a gradation on the lower slopes from the cultivated grassland developed on brown forest soils and humus-iron podzols, map unit 498 (Strichen Association, Class 4) to the improvable grassland on humus-iron podzols, brown forest soils and gleys of the moraines of map unit 503 (Strichen Association, Class 5). Only rough grazing is possible on the peaty podzols and peat of map unit 500 (Strichen Association, Class 6) and on the blanket peat of map unit 4 (Class 6) which dominate the middle and upper slopes above the head dyke. Note the frequent wet flush sites which often coalesce into extensive wet areas. The area in cloud is largely Class 7.* Aerofilms.

Although it is difficult to apply precise altitudinal limits for this unit, it occupies a distinct soil zone dominated by humus-iron podzols. Because the podzol subgroups have been redefined (Handbook 8), many of the iron podzols previously mapped in Aberdeenshire, Kincardineshire, Angus and east Perthshire are now included within this subgroup. Both cultivated and natural soil profiles are found, the latter supporting moorland communities. The herb-rich form of the boreal heather moor community (part of Vaccinio-Ericetum cinereae) is often found on arable land which has been long abandoned and where the soil profile retains evidence of the old plough horizon. Brown forest soils are restricted to certain hill slopes where the effects of birchwood (Trientali-

150

Betuletum pendulae) and colluviation on steep slopes have retarded podzolization. Some peaty podzols are present at higher altitudes. Gleys occupy both small depressions and large areas on concave slopes between interfluves.

On undulating, lowland sites, a moderate range of crops is possible. Within the foothills and the lower slopes of upland areas, additional limitations imposed by climate or pattern, however, restrict the crops to a narrow range. Further hill areas would have been suitable for reclamation to grassland if current trends in land investment had not resulted in their afforestation. Indeed, in recent years, some marginal farms, such as some of those within Glenlivet Forest, have been completely afforested.

Map unit 499   Occurring at higher elevations than *map unit 498*, this unit is found on isolated foothills and on distinct peneplains. It occupies 453 square kilometres or 12 per cent of the association.

The altitudinal range is approximately 300 to 500 metres within a zone in which cool, wet climatic conditions prevail and the dominant soil is a peaty podzol. Free-draining soils, with or without an iron pan, characterize strong and steep slopes and contrast with those soils on gentle slopes where poor drainage conditions are present above the iron pan. Humus-iron podzols on valley sides at lower elevations occupy the narrow zone transitional between *map units 498* and 499. Peat and gleys are confined to small depressions. In contrast with the previous unit, a higher proportion of the soils is formed in shattered rock with a variable veneer of morainic drift.

The vegetation is generally a form of boreal or Atlantic heather moor (Vaccinio-Ericetum cinereae and Carici binervis-Ericetum cinereae) on the freely drained podzols. On the peaty gleys are bog heather moors (Narthecio-Ericetum tetralicis) whereas the peat soils support lowland and upland blanket bogs (parts of Erico-Sphagnetum papillosi).

Soil limitations include the thickness of the peaty surface horizon and shallowness due to shattered rock or to the presence of an iron pan. Although surface seeding is possible on lower slopes, the degree of improvement is limited and the liability to excessive damage from poaching will remain unless drainage is carried out. Over much of the higher moorland sites, climate imposes a further limitation to any improvement. Therefore, much of this unit varies from land suited to improved grassland to rough grazings which have low grazing values. At lower elevations, some land is capable of producing a narrow range of crops.

Map unit 500   In terms of the dominant soil type, landform and vegetation, this unit is similar to the previous one except for the presence of considerable amounts of blanket peat of variable thickness. It occupies 182 square kilometres or 5 per cent of the association. Some humus-iron podzols and peaty gleys are also present; the former are found at lower elevations whereas the latter are often associated with spring lines occurring at a distinct break of slope. The relative proportions of peaty podzols and peat vary widely according to the configuration of the landscape. Shallow, stony drift, or rock debris, occurs on the summits and steep slopes of isolated hills which protrude above the general level of the peat. On exposed, high-level sites, there are subalpine soils, mainly podzols. Till deposits are found on the more gently sloping, lower valley sites, and shallow drift on slopes where restricted rock outcrops occur.

With the majority of the soils having a peaty surface horizon, there tends to be little variation in vegetation, moorland communities being dominant. A few areas, including those previously improved by drainage systems, are marginally suited to improved grassland. Elsewhere, the peaty surface horizon and the climate restrict the land use to forestry and rough grazing. These grazings have low values.

Map unit 501   This minor unit, covering 88 square kilometres (2 per cent of the association) has a scattered distribution throughout central Perthshire, Aberdeenshire and Morayshire. Peaty gleys and peat are the most extensive soils with poorly drained noncalcareous gleys confined to lower elevations and to flushed and receiving sites. Within the undulating lowlands, the topography is similar to that of *map unit 497* but the presence of a peaty surface horizon to the mineral soils is characteristic of this unit. Irrespective of whether the gleys are developed on till or shallow drift, an indurated horizon is commonly found in the profile. At higher elevations within hill masses, peat is widespread, some of it being hagged with long, green ribbons of flushed humic gleys occuring downslope of distinct spring outlets. Peaty podzols are developed on small, scattered, slightly elevated areas with boreal and Atlantic heather moorland communities.

The blanket peat carries lowland and upland blanket bogs (parts of Erico-Sphagnetum papillosi) with heath grass–white bent grassland (part of Junco squarrosi-Festucetum tenuifoliae) in the flushed sites where grazing is particularly heavy.

In lowland and foothill situations, where climatic conditions are not too severe, suitable drainage could permit some reclamation for grassland. At higher elevations, however, the unfavourable effects of soil wetness and peaty surface horizons are compounded by the unfavourable climate, and land use is restricted to rough grazing. Grazing values are low on account of the widespread moorland plant communities.

Map unit 502   Though it occurs on ground steeper than 15 degrees, this unit is closely related to the one previously described. Whereas such steep conditions are widespread on the west coast of Scotland, the unit occurs only around Loch Tay in Eastern Scotland (Sheet 5) and occupies 23 square kilometres (less than 1 per cent of the association).

Map unit 503   Morainic hummocks in valleys, or at the mouths of small glens in lowland areas, are diagnostic of this unit which covers 347 square kilometres (9.0 per cent of the association). Where they occur below about 300 metres in fairly warm, moderately dry or wet climate regions, humus-iron podzols are dominant. Minor areas of brown forest soils are confined to the mounds. Gleys and some peat are found in the associated hollows. The topography is the major influence on the soil pattern. In some cases the mounds are widely spaced and are separated by alluvial flats. Where the mounds extend up the valley sides, they often coalesce to form an irregular, undulating terrain with narrow flushed channels. Surface boulders are locally widespread. The vegetation pattern is extremely variable. In general, the podzols carry boreal heather moors (Vaccinio-Ericetum cinereae) or bent–fescue grasslands (Achilleo-Festucetum tenuifoliae) which are heavily grazed or have been cultivated. Brown forest soils are characterized by acid bent–fescue grassland (part of Achilleo-Festucetum tenuifoliae) with common bracken infestation or birchwood. Rush pastures and sedge mires are present on the poorly drained gleys. Scattered areas of broadleaved woodlands consist mainly of eastern highland oak and birchwood (Trientali-Betuletum pendulae) and Atlantic oakwood and western highland oakwood (Blechno-Quercetum).

Slope and soil pattern are the main factors affecting land use. In restricted areas where the mounds tend to coalesce, the land is capable of producing a narrow range of crops but, elsewhere, the limitations are sufficiently severe to restrict improvement. Large-scale reclamation is often impossible because of the complex topography and the small size of the individual mounds, but some progressive farmers have achieved some grassland improvement. A few areas capable of grassland improvement have been afforested.

Map unit 504   This unit covers 112 square kilometres or 3 per cent of the association. Although there are obvious affinities with the previous unit regarding the parent material and the distribution of morainic hummocks in valley sites, this unit occurs at higher elevations and differs by most soils having a peaty surface horizon. The soil pattern varies according to the relief; peaty podzols are associated with the mounds whereas peat, often 1 to 3 metres thick, and peaty gleys occupy the associated hollows. Freely drained peaty podzols, imperfectly drained peaty podzols and peaty podzols gleyed above an iron pan all occur but have been grouped together. Humus-iron podzols are present at lower elevations and subalpine soils occur on very localized, highly exposed and elevated sites. In narrow valleys, the hummocks are normally in close proximity, whereas in basins the proportion of peat is high and peaty alluvial soils and mineral alluvial fans may occupy relatively significant areas between the hummocks. In both situations, the size and shape of the mounds vary and, where the coarser-textured deposits extend up the valley side, gullying is common. Many of the mounds are rock-cored and many have a bouldery surface.

Moorland communities are extensive with boreal and Atlantic heather moors (Vaccinio-Ericetum cinereae and Carici binervis-Ericetum cinereae) on the peaty podzols and bog heather moors (Narthecio-Ericetum tetralicis) on the wet, peaty soils of the depressions.

Unfavourable climatic conditions restrict the better soils on strong slopes to a marginal capability for improved grassland. In most situations, the high water-retention capacity of the peaty surface horizon, the steepness of the hummocks and presence of boulders prevent mechanized improvement. Most of the land consists of rough grazings with the dominant wet, moist and dry moorland vegetation providing low grazing values. Very severe wetness and trafficability limitations apply to the areas of peat.

Map unit 505   Developed on slopes created by glacial erosion, this unit is confined mainly to central Perthshire, within the valleys of the Rivers Tummel and Tay where it covers 237 square kilometres (6 per cent of the association). Although precipitous rock faces and small areas of non-rocky, but rock-controlled slopes occur, the topography consists mainly of steep, irregular, moderately rocky slopes. The parent material of most of the soils is coarse textured and stony, though relatively stone-free, colluvial material is found on the steeper slopes. The soils are almost entirely free-draining and consist of brown forest soils; some humus-iron podzols occur at higher elevations where the ground vegetation of bent–fescue grasslands (Achilleo-Festucetum tenuifoliae) or eastern highland heathy birchwood (part of Trientali-Betuletum pendulae) is replaced by boreal heather moors (Vaccinio-Ericetum cinereae). Brown rankers are associated with areas of outcropping rock. Poorly drained gleys are confined to distinct and narrow drainage channels.

Outcropping rock and steep slopes make arable cropping impractical and the less steep, rock-controlled terrain is only marginally suited for grassland reclamation. Most of the unit is restricted to rough grazing which is of moderate value. A considerable acreage has been afforested in recent years.

Map unit 506   Whereas the topography is very similar to that of the previous unit, this map unit is widespread on hilly terrain and most of the soils are podzols. It covers 418 square kilometres or 11 per cent of the association. Because of the dominance of podzols, the unit is related to *map units 498* (non-rocky hills) and *509* (very rocky hills) and small patches of the latter two units are included occasionally. The dominant soils form an altitudinal sequence from humus-iron podzols to peaty podzols at about 400 metres. Podzolic and peaty rankers are common, both on the rocky, boulder-strewn hill summits and in the vicinity of rock outcrops. Peat and peaty gleys are confined to natural depressions and to concave slopes within receiving sites. Excluding localized colluvium, the soil parent material is always a shallow drift overlying shattered or solid rock.

Where the soils are podzols, the plant communities are mainly boreal and Atlantic heather moors (Vaccinio-Ericetum cinereae and Carici binervis-Ericetum cinereae). There is some heath rush–fescue grasslands (Junco squarrosi-Festucetum tenuifoliae) at the head of seepage sites within small embayments.

Land use is restricted to rough grazing because of slope, the presence of rock outcrops and the soil pattern. The dominance of moorland plant communities produces grazing of mainly low values though minor areas of moderate value do occur.

Map unit 507   This minor unit is characterized by irregular slopes which have been severely eroded and are moderately rocky. It covers only 151 square kilometres (4 per cent of the association). Although peaty podzols and peaty rankers occur on the rocky knolls, the peaty gleys and peat associated with the hollows are the most extensive soils. The unit is confined to hill and upland regions and is often associated with spring lines. Soils tend to be shallow, but because of active soil creep between rock outcrops, soil depth may vary over short distances. The considerable variation in topography and soil drainage leads to a complex soil pattern and to a wide variation of plant communities. On the peat and peaty gleys, the vegetation is dominated by bog heather moors (Narthecio-Ericetum tetralicis) and upland blanket bog (part of Erico-Sphagnetum papillosi). Heather moorland communities (Carici binervis-Ericetum cinereae and Vaccinio-Ericetum cinereae) are widespread on the peaty podzols and peaty rankers.

The unit is restricted to rough grazing and forestry with little improvement possible because of the peaty surface horizons and the rock outcrops.

Map unit 508   On the steep, rugged slopes alongside Loch Tay and north of Dunkeld, a unit of very restricted extent occurs. It is clearly related to *map unit 505* but is significantly rockier with frequent, precipitous cliffs and some scree. The unit covers only 10 square kilometres (less than 1 per cent of the association). Because of severe glaciation, the mantle of soil parent material is extremely thin and brown rankers and lithosols are common; brown forest soils occur in localized patches of deeper colluvium. Gleys are infrequent. On the brown mineral soils is found a wide range of woodland communities, such as oak, hazel and birchwood which contrasts with the boreal heather moors (Vaccinio-Ericetum cinereae) at higher elevation where humus-iron podzols occur.

Improvement measures are impractical owing to the abundance of rock, steep slopes and shallow soils. Except for the steeper rocky faces, the land is used for forestry and rough grazing, the grazing being of moderate value.

Map unit 509   Complex, very rocky and steep slopes are characteristic of this minor unit which covers 61 square kilometres (2 per cent of the association). It is found generally intermingled with *map unit 506* from which it differs by having a greater proportion of outcropping rock. Excepting that rankers and lithosols are the dominant soils and that precipitous rock faces with active and stabilized scree are common, the other soil, site and vegetation characteristics are similar to those of *map unit 506*.

The land use on this unit is restricted to rough grazing, the plant communities providing grazing of low and moderate values.

Map unit 512   This small unit comprises subalpine soils which are confined to severely eroded terrain with widely contrasting slopes and varying amounts of rock outcrop. It covers 144 square kilometres or 4 per cent of the association, mainly between 550 and 750 metres in the cold, wet uplands of western Perthshire. Exposed mountain summits, where soils are developed on shallow drift and may be subjected to frost-heave, steep-faced corrie walls with considerable rock outcrop, scree and colluvial deposits are the main topographic elements. On the summits, there are minor areas of alpine soils with peat deposits, often eroded, in col sites. On the steeper slopes, noticeable mineralization of the organic surface horizons has taken place and snow-bed plant communities are conspicuous by their green colour.

Land use is restricted to rough grazing. The grazings are of low value over much of the unit although the localized patches of *Nardus* grassland and snow-bed communities provide grazing of moderate values.

Map unit 513   Most of the soil and environmental conditions described for *map unit 512* are found in this unit which is confined to undulating plateaux and hill summits. Extremely hagged peat blankets the landscape, with subalpine and alpine soils restricted to summits and upper convex slopes. The ratio of peat to mineral soil is highly variable. Rankers are of limited occurrence. The unit covers 169 square kilometres or 4 per cent of the association.

Within this environment there is a particularly close relationship between the vegetation and the soil type. The peat carries mountain blanket bog (Rhytidia-delpho-Sphagnetum fusci) where species of *Sphagnum* are locally dominant and the cloudberry (*Rubus chamaemorus*) is a characteristic plant. Some lichen-rich boreal heather moor (part of Vaccinio-Ericetum cinereae) and, more commonly, stiff sedge–fescue grasslands (part of the *Carex bigelowii–Festuca vivipara* Association) occurs on the subalpine soils.

Land use is limited to rough grazings, some of a very poor and seasonal nature; the grazings are of a mainly low value, occasionally moderate.

Map unit 514   Located mainly on broad, undulating mountain summits and ridges in the Glen Lyon area, Perthshire, this unit comprises alpine soils which differ in their morphology according to their location and the completeness of vegetation cover. It covers 54 square kilometres or 1 per cent of the association. Free-draining soils always occupy shedding sites. Most receiving sites in corrie situations are occupied by gleys and in cols by hagged peat. Terracettes, stone stripes and other types of patterned ground are widespread and are related to intense frost-action which has also produced a characteristic soil parent material. The dominant plant communities of alpine azalea–lichen heath (Alectorio-Callunetum vulgaris) and stiff sedge–fescue grasslands (the *Carex bigelowii–Festuca vivipara* Association) often have a patchy cover which contrasts with the extensive areas of bare ground and the complete vegetation cover of alpine clubmoss snow-beds (the *Lycopodium alpinum–Nardus stricta* Community) in localized areas.

Land use is mainly restricted to rough grazings; although small areas of moderate value grazings are characteristic of the grassy snow-beds, most of the vegetation affords only grazing of low values.

The more exposed grazings are of a very poor and seasonal nature.

Map unit 515 The dominant soils of this very restricted unit are rankers and lithosols which occupy corrie walls, crags and scree slopes within the same environment as *map unit 514*. Narrow ridge crests, and mountains with extremely stony and bouldery surfaces are also included. Alpine soils, with some subalpine soils, are developed on mountain-top detritus on stable sites. The unit covers only 9 square kilometres or less than 1 per cent of the association.

Land use is restricted entirely to very poor rough grazings; the grazings, apart from the vegetation in localized snow-beds, have low values. Except for localized occurrences of blaeberry and bog whortleberry heath (the *Rhytidiadelphus loreus–Vaccinium myrtillus* Community and the *Racomitrium lanuginosum–Vaccinium uliginosum* Association) on rocky, steep-sided slopes where lithosols and rankers are widespread, the plant communities are similar to those of *map unit 514*.

THE TARVES ASSOCIATION

(Map units 517–518, 520–525, 527, 529, 530, 532–534)

The Tarves Association is one of the more extensive and agriculturally significant associations and covers 1255 square kilometres or 5.1 per cent of the land area in Eastern Scotland (Sheet 5). With a distribution ranging from the lowlands of north-east Scotland to the alpine environment of the mountain summits, a large number of map units, fourteen in all, have been identified. The soils are developed on drifts derived from igneous and metamorphic rocks of intermediate basicity, for example, hornblende-schist, biotite-gneiss and diorite, and from a mixture of acid and basic rocks. The latter drift is the most extensive and consists of till deposits which are characteristic of lower, convex and straight slopes; shallow, stony drifts are associated with upper slopes and summits. The till is generally a yellowish brown colour with a sandy loam or sandy clay loam texture on free-draining sites. In sites with poor drainage, textures of loam or clay loam are more typical. The thickness of till, in both cases, exceeds 1 metre and the surface layers may be water-sorted. In comparison, the drift deposits on the upper hill slopes and summits are stonier and coarser in texture and therefore have some affinity with the shallow drifts on shattered rock mapped within many other associations, for example, the Strichen Association. In such situations, the presence of shattered rock within profile depth is not usually evident except where the soils are formed *in situ* from rocks of intermediate basicity. In certain localities, the rock is deeply weathered. Other parent materials include mountain-top detritus and coarse-textured drifts deposited as moraines and colluvium. Very restricted soils developed on parent material formed from mainly acid rocks, but flushed by drainage water from limestone, have also been assigned to this association.

The association occupies a broad band of variable terrain stretching north-eastwards from the Tarland basin in Aberdeenshire to within a few kilometres of the coast between Aberdeen and Peterhead. Further extensive areas occur in the upper reaches of the River Don and along the valley of the River Deveron between Bridgend and Milltown. Smaller areas are located in most of the Angus glens, in eastern Perthshire and in the foothills and uplands extending south-westwards into the Trossachs.

Because of the wide altitudinal limits of the association, a range of soils exists. owing to the relatively base-rich parent material, there is a higher proportion of brown forest soils, both freely and imperfectly drained, than in associations formed on more acid parent materials, for example, Countesswells and Strichen Associations. Such soils are characteristic of the lowlands and of the lower slopes of valleys which extend into the foothills and where humus-iron podzols might be expected. Profiles in the latter sites often have a faint eluvial horizon or abundant bleached sand grains beneath a moder horizon, suggesting that the former is a horizon transitional to the E horizon of a podzol. Humus-iron podzols are present, however, at higher altitudes, mainly under moors but occasionally under planted coniferous woodland in Deeside and Donside. With increasing altitude, humus-iron podzols merge into peaty podzols and, in upland regions, subalpine and alpine soils occur. Many of the peaty and noncalcareous gleys are surface-water gleys in which the effects of poor drainage are confined to the top metre because of the presence of induration or fine-textured subsoil horizons.

With most of the association devoted to agriculture, arable and permanent pasture communities (Lolio-Cynosuretum) are extensive. Acid bent–fescue grassland (part of Achilleo-Festucetum tenuifoliae) is widespread on freely drained brown forest soils and podzols, especially on land which has been cultivated or which is heavily grazed. Dry and herb-rich boreal heather moors (parts of Vaccinio-Ericetum cinereae) cover most of the podzols. The potential of the high base saturation tends to be lost at higher elevations where climatic conditions promote peat development. Lichen-rich boreal heather moor (part of Vaccinio-Ericetum cinereae), alpine azalea–lichen heath (Alectorio-Callunetum vulgaris) and stiff sedge–fescue grasslands (the *Carex bigelowii–Festuca vivipara* Association) are common within the higher zones, where subalpine and alpine soils are dominant.

The free-draining brown forest soils occur in the climate regions of warm, wet lowlands or in the more exposed areas of warm, moderately dry lowlands where climatic conditions restrict intensive arable farming. However, most of these areas, together with those of the imperfectly drained soils, produce a moderate range of crops. In the latter areas, slow subsoil permeability and problems relating to the timing of cultivation and harvest are additional limitations. Where slightly wetter conditions prevail in the foothills, the land is marginal though many good-quality stock-rearing farms are located within these areas. Some of the lower valley sides, where climate is not a serious limiting factor, are suited to grassland reclamation, for example, much of upper Donside and the Angus glens. At higher elevations, soil pattern, occasional boulders and outcropping rock prohibit grassland improvement though the inherent moderate grazing values play an important role in the local agricultural economy. Land use in the very exposed mountain summits and areas of hagged peat within the alpine region is restricted to seasonal rough grazings which are mainly of low value.

Map unit 517   This unit is the most extensive within the association (630 square kilometres or 50 per cent of the association) and occupies large tracts of lowland and low hill areas in north-east Scotland as well as valleys within the foothills, such as those in the Strathdon and Ballater areas and the middle and upper reaches of the River Deveron. In these areas, local intrusions of basic igneous rocks have contributed to a mixed till.

The dominant soils are freely and imperfectly drained brown forest soils, most of which are cultivated with topsoils about 25 centimetres thick except where cultivation has been long established, for example, near Oldmeldrum where topsoils are up to 75 centimetres thick. In addition to a variation in texture from sandy loam to sandy clay loam related to the topographic position, the thickness of the till also varies. On convex slopes, where the freely drained soils are developed in a sandy loam till, the underlying rock is usually less than 1.25 metres deep. In lowland sites, however, imperfectly drained soils are developed on a sandy clay loam till which is generally much thicker. Gleys are of minor extent and are usually developed on the sandy clay loam till on concave slopes or within depressions. Humus-iron podzols are found on the strong and steep slopes of some hill sides where rock outcrops and very shallow parent material have prevented arable agriculture. Though the dominant soils possess no obvious limitation to arable agriculture, moderate climate limitations make careful management essential. Only in more sheltered areas is the land capable of producing a wide range of crops. Elsewhere, the soils are capable of producing only a moderate range.

Map unit 518   This extensive unit occupies 21 per cent of the association (263 square kilometres) and comprises noncalcareous and peaty gleys with minor humic gleys and peat. It is distributed widely throughout the undulating lowlands of north-east Scotland on footslopes and in distinct depressions where ground-water gleys are developed. On the valley sides within the foothills, for example, the Deveron valley and Glen Prosen, the unit characterizes concave slopes where both surface-water and ground-water gleys are formed in compact, sandy clay loam till. Such soils also occupy distinct embayments, for example, within Glen Shee and Glen Prosen where there is intense flushing. In these situations, the soils are developed on a moderately coarse-textured, sandy loam drift which is often water-modified and indurated.

On the flushed slopes, the white bent–tussock-grass grassland (the *Cirsium palustre–Nardus stricta* Community) is the dominant plant community.

Within both the valley and embayment sites, there are also low mounds with cultivated brown forest soils which often support acid bent–fescue grassland (part of Achilleo-Festucetum tenuifoliae) sometimes with bracken infestation, or common white bent grassland (part of Junco squarrosi-Festucetum tenuifoliae). Under semi-natural vegetation, these soils have a weakly developed E horizon.

The limitations associated with poor drainage can be ameliorated by intensive tile drainage, except in depressions with a high ground-water table where the soil remains too wet to permit tillage. Therefore, within the lowland area much of this unit is capable of producing both a moderate and narrow range of crops. However, in some valley sites around 300 metres, the limitations associated with poor drainage, for example, prolonged waterlogging, timing difficulties in planting and harvest operations, are so compounded by unfavourable weather that cultivation is not feasible. Improved grassland is the most productive form of land use in these situations, except where the soil pattern and surface boulders constitute additional limitations.

Map unit 520   This unit occupies 104 square kilometres (8 per cent of the association) on upper slopes within the undulating lowlands and foothills where humus-iron podzols are dominant, for example, the Strathdon area. In hilly areas, it occupies the strong and steep slopes above the general limit of arable agriculture. The soil parent material is a stony, sandy loam till, often less than a metre thick, and with a strong platy structure in the indurated subsoil. Plant communities are dominated by dry boreal heather moor (part of Vaccinio-Ericetum cinereae) or acid bent–fescue grassland (part of Achilleo-Festucetum tenuifoliae) where cultivation has been abandoned or intensive grazing is practised. Small areas of gleys are associated with rush pastures.

Soil and topographic limitations combine to restrict arable agriculture to a narrow range of crops and most of the unit has potential only for improved grassland. A considerable proportion of the unit has undergone afforestation in recent years.

Map unit 521   Occupying 79 square kilometres (6 per cent of the association) on hill and valley slopes between 300 and 500 metres, this unit comprises peaty podzols, with some humus-iron podzols on steeper slopes and in afforested areas. It is extensive in upper Glen Prosen, upper Donside and in the Ballater area. In most instances, the parent material is a shallow, sandy loam drift and the landscape slightly rocky. An iron pan is only locally developed; dependent upon local variation in slope, the drainage class above the pan varies from free to poor. On narrow cols and flat summits, there are peat deposits which, together with gleys, form minor soil components of this unit. Moorland communities are the most extensive, especially dry and moist boreal heather moors (parts of Vaccinio-Ericetum cinereae); common white bent grassland (part of Junco squarrosi-Festucetum tenuifoliae) is widespread on steep slopes where flushing is active and where there is intensive grazing.

The greater part of this unit is restricted to rough grazings of moderate and low values.

Map unit 522   Occurring in Strathdon and in small, scattered localities throughout the eastern Grampian Highlands, this minor unit has definite affinities with the previous unit, but differs by possessing a higher proportion of dystrophic peat. The ratio of peat to peaty podzol varies according to the configuration of the landscape, with peat occupying cols, saddles, small basins and gently sloping summits. On some very exposed hill summits at high elevations, subalpine soils are developed. Peaty gleys are locally extensive in the vicinity of spring lines at distinct breaks of slope. Lowland and upland blanket bogs (parts of Erico-Sphagnetum papillosi) are dominant in situations where peat and peaty gleys are found. Soil and vegetation characteristics are otherwise very similar to those of the previous unit. It covers 60 square kilometres (5 per cent of the association).

Although most of the unit is restricted to rough grazing with moderate and low grazing value, there is potential for grassland reclamation in areas of more favourable climate and terrain. Because all the soils have a peaty surface horizon, reclaimed areas must be utilized with care to prevent severe poaching damage.

Map unit 523   This restricted unit occupies only 13 square kilometres (1 per cent of the association) and occurs in high-level embayment sites and smooth hillslopes between 350 and 600 metres in Glens Tanar, Prosen and Shee. On concave slopes which act as receiving sites for the headwaters of streams, peat and gleys are dominant with surface horizons often flushed. Humic gleys are widespread in long flushed ribbons where they support common sedge flushes (the *Carex nigra* Community) that are distinctive and contrast with the surrounding dominant bog heather moors (Narthecio-Ericetum tetralicis) and blanket bogs (Erico-Sphagnetum papillosi).

Land use is limited mainly to rough grazings which have low grazing values. The long flush ribbons have a higher grazing potential and act as sheep runs between this zone and the better, high-level grazings.

Map unit 524 This very restricted unit is confined to 5 square kilometres (less than 1 per cent of the association) of hummocky topography on the valley floors and lower slopes of Glens Callater and Doll. Because of the altitude, between 500 and 600 metres, peaty podzols are dominant on the mounds, with dystrophic peat or peaty gleys in the associated hollows. In limited areas, where the coarse-textured parent material is enriched with limestone from narrow out-cropping bands, the soil pattern is dominated by brown forest soils and non-calcareous gleys. The high base saturation and pH values tend to suppress podzolization and the characteristic communities of boreal heather moors (Vaccinio-Ericetum cinereae) and bog heather moors (Narthecio-Ericetum tetralicis) are replaced respectively by bent–fescue grasslands (Achilleo-Festu-cetum tenuifoliae) and soft rush pastures (the *Ranunculus repens–Juncus effusus* Community). Blanket bogs (Erico-Sphagnetum papillosi) are the characteristic communities on the dystrophic peats of the hollows.

The localized areas of free-draining, mineral soils can be mechanically improved for grassland except where the slope restricts the use of machinery. Land use is limited mainly to rough grazing; the grazing values are low and moderate.

Map unit 525 Brown forest soils dominate this minor map unit which is confined to small areas of Glens Prosen and Clova and to hillslopes near Aberfeldy. They cover 13 square kilometres (1 per cent of the association). On moderately rocky, steep slopes, the soils are derived often from colluvium with deposits of shallow, stony drift confined to the more stable sites. Beneath precipitous rock faces there are often stabilized screes, and on irregular slopes, close to outcropping rock, brown rankers and lithosols are common. Humus-iron podzols are restricted to sites of higher elevation where the herb-rich boreal heather moor (part of Vaccinio-Ericetum cinereae) is widespread. In other situations, bent–fescue grasslands (Achilleo-Festucetum tenuifoliae) and eastern highland oak and birchwood (Trientali-Betuletum pendulae) are dominant, the former often with a bracken infestation. Minor patches of boreal juniper scrub (Trientali-Juniperetum communis) are associated locally with the steep slopes. High microbial activity occurs beneath such communities on the steep slopes where surface organic horizons are often heavily mineralized owing to colluviation. With extensive reclamation being impractical because of the steep slopes, outcropping rock and surface boulders, the land is used mainly as rough grazing; the grazing values are mostly moderate.

Map unit 527 Confined to small areas within upper Deeside and the Angus glens, this map unit possesses landforms and soil parent materials similar to those of the previous unit. It covers 11 square kilometres (less than 1 per cent of the association). However, humus-iron and peaty podzols are the principal soils, with dry boreal heather moor (part of Vaccinio-Ericetum cinereae) the dominant plant community. High-quality grazing on heath grass-white bent grassland (part of Junco squarrosi-Festucetum tenuifoliae) is confined to the steep, flushed channels where humic gleys or imperfectly drained podzols are locally extensive.

Where the soils tend to be shallow, the influence of the underlying basic rock, for example, hornblende-schist, is sufficient to produce localized patches of herb-rich bent–fescue grassland and upland bent–fescue grassland (parts of Achilleo-Festucetum tenuifoliae) which are heavily grazed.

Although improvement to provide better grazing is feasible on a mosaic pattern, the land is usually maintained as rough grazing. The value of the grazing, and natural shelter for lambing, however, means that such land has great significance within the overall economy of these upland farms.

Map unit 529   The soils of the moderately and very rocky, steep and very steep slopes in Glen Doll and Glen Clova have been included within a very restricted map unit which is related to *map unit 525*. It covers only 3 square kilometres (less than 1 per cent of the association). The underlying geology is complex with out-crops of hornblende-schist and diorite, both of which may be interbedded with acid schist. The soils include brown rankers, and brown forest soils with some gleys and humus-iron podzols. Scree is an important element in the unit. The altitudinal range extends from 350 to 800 metres and peaty rankers and peat are found at the upper limits despite the steep slopes and southerly aspect.

The high basicity of the soils and the constant addition of nutrients through colluviation ensure that the rough grazings are of moderate value. Large areas, however, consist only of rock. The main plant communities are bent–fescue grasslands (Achilleo-Festucetum tenuifoliae) and dry boreal heather moor (part of Vaccinio-Ericetum cinereae). Crested hair-grass grassland (the *Galium verum–Koeleria cristata* Community) is also present.

Map unit 530   Although this map unit is not extensive, covering only 9 square kilometres (less than 1 per cent of the association), it is conspicuous by its rugged appearance within Glen Doll, Glen Clova, the Balmoral area and the head of Glen Prosen. The unit is moderately or very rocky and it occurs on steep or very steep slopes. The soils are mainly rankers and humus-iron podzols. As with *map unit 529*, the parent rock varies from hornblende-schist to diorite and both rock types may be interbedded with limestone, quartzite and acid schists.

Rough grazing, mainly of low value, is provided by the plant communities which include dry boreal heather moor (part of Vaccinio-Ericetum cinereae) and the blaeberry heath (the *Rhytidiadelphus loreus–Vaccinium myrtillus* Com-munity).

Map unit 532   This unit consists of subalpine soils and peat. It comprises mountain summits at the head of Glens Callater, Clova and Shee and within the Strathdon area and covers 28 square kilometres (2 per cent of the association). Because a wide range of rock outcrop and slope class is accommodated within this unit, there is a considerable variation in the proportion of minor soil types. They include peaty podzols, peaty gleys, rankers, lithosols and some alpine soils. There is a corresponding mosaic of plant communities. Upland bent–fescue grassland (part of Achilleo-Festucetum tenuifoliae) on steep colluvial and flushed slopes contrasts with the mountain blanket bogs (Rhytidiadelpho-Sphagnetum fusci) on peat deposits and lichen-rich boreal heather moor (part of Vaccinio-Ericetum cinereae) on exposed mountain summits.

Land use is restricted to rough grazing which has low, occasionally moderate, values; very localized communities with high values form important high-level grazings. These relieve the pressure on grazings at lower altitudes during the summer months.

Map unit 533 This unit has been separated within the zones of subalpine and alpine soils to accommodate 34 square kilometres (3 per cent of the association) of mountain summits, upper valleys and cols where peat deposits are extensive and are co-dominant with subalpine and alpine soils. Such areas are located on hills around Strathdon, Ballater and Braemar. Most of the peat is extremely eroded with intense hagging, which in places reveals an iron pan or weathered rock at depth. Where the vegetation is influenced by mineral soil at the base of the hags, green ribbons of stiff sedge–fescue grassland (the *Carex bigelowii–Festuca vivipara* Association) are conspicuous within the dominant mountain blanket bogs (Rhytidiadelpho-Sphagnetum fusci). Depressed sites with subalpine gleys are normally heavily flushed by springs.

Land use is limited to rough grazings; the grazings are mainly of low value, with some of a very poor and seasonal nature.

Map unit 534 Occupying the summits above about 800 metres, the unit consists of soils, plant communities and associated features which are characteristic of the extremely severe environment of the alpine zone. It covers 5 square kilometres (less than 1 per cent of the association area). Such terrain is extremely limited within this association and is confined to individual hill summits such as The Buck (Cabrach), Little Glas Maol (Glenshee) and Mayar (Glen Prosen). Most slopes are gentle or strong, except in the vicinity of local rocky knolls where steep slopes may occur. Rock outcrops are often in the form of tors.

Climatic limitations are extremely severe and restrict this unit to very poor rough grazings of a seasonal nature.

The dominant plant communities are alpine azalea–lichen heath (Alectorio-Callunetum vulgaris) and stiff sedge–fescue grasslands (the *Carex bigelowii–Festuca vivipara* Association). Mountain blanket bog (Rhytidiadelpho-Sphagnetum fusci) characterizes the minor areas of peat.

THE TIPPERTY/CARDEN ASSOCIATIONS

(Map unit 545)

The soils of the two associations comprising this group are developed on lacustrine silts and clays derived from Old Red Sandstone sediments. Together, they occupy 51 square kilometres (0.2 per cent of the land area). The Tipperty Association is found at altitudes below 75 metres and the main area, approximately 40 square kilometres, occurs to the south of Peterhead and extends along the coastal plain to the north-east of Ellon. The most southerly area of the association occurs about 6 kilometres north of Aberdeen and includes the Tipperty clay pit. The soil parent material consists of red clays and silts which are calcareous below 1 metre and have occasional thin bands of sand.

Soils of the Carden Association, covering a total area of 6 square kilometres, occur at altitudes up to 45 metres and are found in four small areas to the west of Elgin. The parent material is a reddish silty clay which is well bedded with occasional thin layers of coarser material.

Map unit 545 is the only unit distinguished and consists of brown forest soils with gleying together with some poorly drained noncalcareous gleys and minor areas of poorly and very poorly drained peaty gleys. It occurs on undulating lowlands with gentle slopes. The brown forest soils with gleying normally have a clay loam topsoil with a subangular blocky structure on a clay subsoil with an angular blocky structure varying to medium prismatic and overlying a red clay with a massive structure. In the Carden Association area, however, the topsoils often have a loam texture, the relatively high sand content being due to the incorporation of windblown sand. Though the upper horizons may have a few small pebbles, all other horizons are stone-free. A high base status is common. Most of the unit is arable or permanent pasture (Lolio-Cynosuretum). Areas of peaty gley support rush pastures and sedge mires.

To the west of Elgin, the unit occurs in a climate region of warm, moderately dry lowlands with an average annual rainfall of 650 millimetres. The land is highly productive, producing a wide range of crops with high yields. South of Peterhead, the climate region is one of fairly warm, moderately dry lowlands and foothills with an average annual rainfall of 760 millimetres. The soils there have a slightly higher clay content which creates workability problems. The land is used mainly for arable agriculture and produces a moderate range of crops with high or average yields.

THE TYNET ASSOCIATION

(Map units 565–567)

The parent materials of this association consist of drifts derived from sandstones and conglomerates of Middle Old Red Sandstone age. It is normally red or reddish brown, with a sandy loam or loam texture which often becomes finer with depth. The soils are always stony with rounded pebbles of quartzite, quartz-schist and granite acquired from the conglomerate. Angular fragments of sandstone and fine conglomerate are also found. Podzols dominate the association. Freely drained humus-iron podzols are the most extensive, with imperfectly drained humus-iron podzols being locally important. Peaty podzols are confined mainly to hills above 150 metres. Noncalcareous, humic and peaty gleys are minor components. An indurated layer is usually present in most soils.

The association covers 44 square kilometres, or 0.2 per cent of the land area, and stretches from Buckie south-westwards to the River Spey valley, its main location being on the foothills which border the Moray coastal lowlands. Rock outcrops are rare and slopes are usually gentle or strong, locally steep. A widespread feature is deep gullying of the soft parent rocks which underlie the drift. Altitude ranges from 30 to 270 metres.

Climate regions include warm, moderately dry lowlands and fairly warm moderately dry lowlands and foothills. The average annual rainfall is between 700 and 900 millimetres.

Forestry and agriculture are the main forms of land use with forestry being dominant. The main factors affecting agricultural capability are stoniness and induration.

Map unit 565   This unit covers 1 square kilometre (less than 5 per cent of the association) in four small areas north-west of Keith. They occur on gently sloping, receiving sites where noncalcareous gleys are the dominant soils. Minor patches of humic gleys, peaty gleys and peat are included.

Most of the unit is afforested. Wetness and to a lesser extent, climate are the main factors affecting the agricultural use of the land.

Map unit 566   Covering 30 square kilometres, this unit comprises 70 per cent of the association area. Freely drained humus-iron podzols, some cultivated, are the dominant soils. Imperfectly drained humus-iron podzols occupy some gentle slopes. Peaty podzols, restricted to the higher altitudes, and gleys in receiving sites are minor components.

The landform consists of undulating lowlands and hills with gentle and strong slopes. Most of the lowlands, south of Buckie, are cultivated but as the land rises to the south, afforestation becomes dominant. An area of imperfectly drained humus-iron podzols, north of Mulben, is also cultivated.

The main plant communities, excluding the plantations, are arable and permanent pastures (Lolio-Cynosuretum); dry Atlantic heather moor (part of Carici binervis-Ericetum cinereae) on the uncultivated podzols and rush pastures and sedge mires are minor communities.

Land use is divided almost evenly between arable agriculture and forestry. Stoniness and induration, together with climate in the higher areas, are the major factors limiting the agricultural capability. The low ground is capable of producing a moderate range of crops whereas the high ground is suited only to the production of grass.

Map unit 567 This unit covers 13 square kilometres (30 per cent of the association) on the hills south of Fochabers. It occurs mostly above 150 metres on non-rocky terrain with gentle to steep slopes. Peaty podzols are the dominant soils; humus-iron podzols, both imperfectly and freely drained, are minor components.

Plant communities include boreal and Atlantic heather moors (Vaccinio-Ericetum cinereae and Carici binervis-Ericetum cinereae respectively) though almost all of the unit is afforested. Climate, stoniness and induration are the limiting factors which make this unit only moderately suited to grassland reclamation.

MISCELLANEOUS LAND UNITS

Bare rock, scree and cliffs
This map unit, covering 19 square kilometres, accommodates areas of extremely rocky land in the strongly glaciated mountains of the South-West Grampian Highlands. It is mainly associated with the mountain tops. Elsewhere in Eastern Scotland such areas of rock pavements, cliffs and scree are included in the appropriate map unit dominated by rankers and lithosols.

Built-up areas
Such areas total 208 square kilometres. Most of the major centres occur in the coastal lowlands where they have developed around natural harbours.

Freshwater lochs
The areas of these lochs in Eastern Scotland total 265 square kilometres. For the purpose of calculation of regional area percentages, the area of these lochs has been excluded as part of the total land area.

3 Land Evaluation

Earlier chapters of this book have described the main natural resource attributes of Eastern Scotland (climate, landform, soil and vegetation) and classified them into a number of units. The characteristics of each of these units influence Man's use of the land contained within it. Land evaluation is the assessment of a range of possible uses of the land units, for example, for agriculture, forestry, recreation or engineering. It incorporates not only the physical attributes of the land but also Man's resources of technology, finance and labour. Because these are variable with time in a manner not accurately predictable, systems of assessing the capability of land for any specific purpose usually attempt their standardization. The potential use of the land then may be assessed under the standard conditions and expressed as capability classes. Land evaluation is not static but must be reviewed periodically and repeated when significant changes take place in any of the human resources.

Land capability classifications are not recommendations for the particular use of a piece of land. Their purpose is to identify areas where that use may be carried out most easily. Only by carefully comparing all the alternatives and incorporating economic and political judgements, in particular cases, can recommendations for actual land use be determined. For this reason no one map indicating 'best land use' is likely to be achieved.

In Scotland, a system of land capability classification for general agricultural purposes has been constructed (Bibby *et al.*, 1982). An explanation of its broad principles and the parameters used in its application in Eastern Scotland form the bulk of this chapter. The final section provides some comments on the effects of natural resources on other uses for which fuller classification systems have not been constructed yet.

LAND CAPABILITY CLASSIFICATION FOR AGRICULTURE

The land capability classification for agriculture has as its objective the presentation of detailed information on soil, climate and relief in a form which will be of value to land-use planners, agricultural advisers, farmers and others involved in optimizing the use of land resources.

Its applications include the following:

1 Contributing to an inventory of the national land resource
2 Providing a means of assessing the value to agriculture of land on a uniform basis as an input to planning decisions
3 Defining major limitations to land use
4 Assisting in environmental and amenity planning
5 Contributing to farm and estate planning and to technical advisory work

PHYSICAL FACTORS AND THEIR EFFECT UPON AGRICULTURE IN EASTERN SCOTLAND

Climate With Eastern Scotland including the highest, single mountain mass in the country, and ranging from the main watershed along its western boundary to sea level, the climate is variable and often extreme. The principal trends, reflecting the dominating influence of the Grampian Mountains, are the falling temperature and increasing exposure which accompanying rises in altitude, and the easterly decreasing rainfall. Orographic rainfall in the west and the resultant föhn winds help to produce relatively dry and warm conditions over much of the sheltered, eastern lowlands. There, the climatic constraints are minor and moderate, and with a rainfall mostly below 900 millimetres arable cropping is widespread. In some coastal areas, particularly in Morayshire, the rain-shadow effect is pronounced and drought problems result when the low rainfall, often less than 700 millimetres, coincides with coarse-textured soils. Around east Fife and in parts of Strathmore, the highly favourable climate has encouraged the development of soft fruit and vegetable field crops. Only in the Forth Lowlands, where the rain-bearing westerlies have easy access to the low ground, does rainfall restrict agriculture to grassland systems.

In the foothills fringing the Grampian Highlands and Northern Highlands, the decreasing temperatures and shorter growing season, together with increasing exposure, progressively restrict arable cropping, and farming enterprises become based on stock-rearing and breeding. Cereals and root crops, however, are often favoured in the drier areas in the north-east despite the problems associated with sowing and harvesting. With increasing altitude, the climatic factors become more restrictive until all systems are based on grassland; reclamation for grazing alone seldom exceeds 400 metres. Across the mountains, the very severe and extremely severe climatic constraints limit agriculture to rough grazing with sheep frequently replaced by sporting activities centred on grouse-moors and deer-forests.

Gradient Consisting mainly of undulating plains, fluvioglacial terraces and raised beaches, the lowlands have slopes usually much less than 11 degrees and gradient is seldom a serious limitation to arable cropping. Only on slopes above that magnitude are difficulties encountered by machinery such as combine harvesters and forage harvesters. In the foothills and in certain lowland situations, however, slope and configuration, often related to pattern, pose greater problems, for example, in hummocky moraines in the South-West Grampian Highlands and in drumlin fields in the upper Forth valley. Gradient, as well as inhibiting arable cropping, frequently limits reclamation and its subsequent maintenance. Where reclamation to grassland for grazing without conservation is the objective, operations up to 30 degrees are theoretically within the capability of four-wheeled drive tractors. In practice, nature of slope and surface often restrict the potential. Wet surfaces, type of hitch, system of braking, problems associated with turning and weight redistribution following spreading operations are all limiting factors. Although four-wheeled drive tractors and expensive machinery are often not justifiable in marginal farming, reclamation projects are being tackled increasingly by specialist firms using tracked machinery. Even then, gradient restricts operations at around 30 degrees, with rafting on peaty surfaces, especially on those overlying induration, a serious hazard.

Soil   Because of such varied climatic, geological and topographical conditions in Eastern Scotland, soil parent materials are equally variable. They encompass aeolian sands, fluvioglacial, morainic and marine deposits, coarse- to fine-textured tills, locally derived drift and weathered rock. The derived soils have different inherent fertility and range from single-grain sands to massive clays; they may be qualified also by stoniness, induration, shallowness and droughtiness. All these attributes variously affect agricultural capability.

Most of the soils of the Moray Firth Lowlands and some in the North-East Lowlands have inherently low fertility and are moderately coarse textured; many are stony, shallow due to induration and suffer from trace element deficiencies, especially copper and cobalt. Raised beach soils and fluvioglacial soils there, and elsewhere in Eastern Scotland, are restricted by coarse textures, weak structures and are often stony: they suffer similar severe trace element deficiencies except where there is an appreciable argillaceous content to the soil. All these soils are easily worked but have poor nutrient- and water-retention capacities. In stony soils the stone content contributes more to wear of farm implements than does shear strength or sand content and can severely affect the performance of or cause damage to such machines as thinners and precision seeders.

Other soils in the North-East and Central Lowlands are moderately fine and fine textured, and are often inherently fertile. Most of the former are usually only imperfectly drained because of a well-developed blocky structure in the upper horizons which are often water-worked and coarser textured. In the Carse of Stirling, southern Fife and parts of Buchan, however, the fine-textured soils usually have massive subsoils and a low permeability. The narrow moisture range between their field capacity and upper plastic limit results in severe workability problems in wet weather; capping is particularly severe where silt contents are high. Conversely, the high shear strength of the clayey soils creates difficulties in tilth production during a dry cycle. Thixotropy is often a major problem in drainage works where the soils are bedded and well sorted into uniform particle sizes, causing severe slumping in ditches and open drains.

Wetness   Soil wetness reacts with different soil properties and topography to affect farming through workability, trafficability and liability to poaching. In Eastern Scotland, much of the lowlands comprises freely drained and imperfectly drained soils with moderately coarse and moderately fine textures subject to less than 900 millimetres rainfall, and there are few wetness limitations. Where soils with these textures occur in depressions and are subject to ground-water gleying, however, field capacity is reached rapidly in wet weather and soil wetness becomes a serious problem. In Fife, where the rainfall is slightly higher, the low permeability of fine-textured soils, allied to their coarse prismatic or massive structure, frequently induces surface waterlogging; drainage systems are only partially effective. As rainfall increases westwards across the Forth Lowlands to exceed 1600 millimetres, nearly all the soils are likely to remain at field capacity for long periods and be liable to poaching and trafficability problems; the workability of the finer-textured soils is sharply reduced by the increased plasticity.

In the mountains and foothills, most parent materials are permeable and relatively shallow but their permeable horizons are frequently truncated by induration and iron pans. These impermeable horizons often induce surface-water gleying in high rainfall areas and elsewhere in gentle slopes below spring lines, the latter usually occurring at the junction of concave till slopes with convex upper rock-controlled slopes. Most of these soils have organic surface horizons which form in conditions ranging from anaerobic waterlogged to aerobic freely drained, the former associated with gleys and peaty gleyed podzols and the latter with humus-iron podzols and peaty podzols. Usually those of the gleys remain wet throughout the year and are only marginally suitable for reclamation. On planation surfaces in the Highlands, the climate and slow run-off favour the development of blanket peat which also remains wet throughout the year except where hagging increases the percolation rate; only basin peats in the lowlands have responded to reclamation.

Flooding is a serious problem in Highland valleys resulting from both rainstorms and melting snow especially in major rivers such as the Rivers Spey and Findhorn where the tributaries have a rapid fall. Varying from flash-flooding in narrow valleys to floods lasting several days in the wider straths, the more devastating effects have been largely reduced by rock embankments: where flooding risks are high farming is normally restricted to long ley pasture.

Erosion Erosion is a comprehensive term applied to the various ways in which natural agencies obtain and remove soil particles. In Eastern Scotland, the agencies include wind, water, both fresh water and wave action, and gravity. Good land management can often reduce the effects by judicious land use.

Wind erosion is a potential hazard in all the coarse-textured soils in the low rainfall areas and is particularly severe along the southern Moray Firth Lowlands in periods of drought. There, the prevailing south-westerlies in springtime often coincide with maximum degradation of the widespread, weakly developed soil structures during seed-bed preparation. Windblow can result occasionally in blocked roads; sand scorching of seedlings and their complete removal, necessitating re-sowing, are common. Major wind erosion has always been associated with the sandy raised beaches whose evolution has often created large areas of dunes; centuries ago, at Culbin and Forvie, mobile sands overwhelmed settlements. Nowadays, stabilization by afforestation, the creation of golf courses and thatching by marram grass have halted or greatly reduced movement, though such areas are still in a delicate balance. Similar but small-scale problems exist on some loch beaches, for example, Lochs Morlich and Laggan.

Mass wasting by gravity is almost imperceptible in the short term but produces recognizable and important consequences. Creep on slopes, accelerated by downslope ploughing, has resulted in differences in topsoil thickness of moderately coarse-textured soils of more than 0.5 metres in a single field. Natural creep on hill slopes often inhibits development of a peaty surface and induces plant communities with high and moderate grazing values atypical of the surrounding areas. On the mountain tops, gravity, combined with frost-heave and wind erosion, produces characteristic terracettes and other erosional features which usually substantially reduce the already scanty plant cover.

Because much of the coast is non-rocky and still evolving, deposition and erosion are concurrent processes. Wave action is often severe, causing marked loss, for example, in Burghead Bay and Culbin, and necessitating massive sea-wall defences as at Aberdeen. Generally, the land is of low agricultural value but high amenity value. Erosion of valuable alluvium in many river catchments has been occasionally severe but is now largely contained by rock embankments; a notable exception includes the lower reaches of the River Spey below Fochabers which are constantly braiding and prevent any agricultural use of the lowest terrace.

In the hill land, erosion is not a serious problem and is usually localized, often resulting from badly constructed hill roads with inadequate drains and culverts and by the leaders of hill drains debouching on to open slopes. Minor landslips, outwith natural gullying, occasionally follow when road side-cuts are made in peaty gleys with indurated subsoils and in peaty podzols gleyed above an iron pan where gravity, the truncated rooting system and the increased lubrication combine to induce shearing. On the plateaux above 500 metres, however, erosion is widespread on the blanket peatlands and is gradually reducing the area of vegetated surfaces. The subsequent drying out of the hags causes the replacement of higher plants by mosses and lichens and a reduction in grazing values. It is possible that this erosion will accelerate and alter the run-off rates of the catchments.

Pattern   This is not a serious limitation to arable cropping in Eastern Scotland and, throughout the region, is related largely to moundy fluvioglacial and morainic deposits; the latter are largely a feature of the South-West Grampian Highlands. In addition, some of the raised beach areas also display rapid lateral variation in both texture and drainage; for example, where excessively drained sand dunes are superimposed on poorly drained sands and basin peats they are of little agricultural value, and mainly afforested.

The moundy areas of sands and gravels frequently have wet hollows, sometimes peat-filled, which are difficult or impossible to drain. The steep and rapid changes in gradient create problems in the deployment of machinery, and erosion of the crests is increased by cultivation. Droughtiness on the shallow and loose topsoils becomes excessive and leads to premature ripening and substantial reduction in cereal yields. Similar moundy topography occurs in some moderately coarse-textured tills along the Highland Boundary Fault, especially around Callander, but there droughtiness is not a major accompanying limitation. Except for strongly glaciated areas mainly in the South-West Grampian Highlands, most hill slopes have uniform soil conditions with rock outcrops largely restricted to rock-dominated summits in areas naturally limited by climate to rough grazing. On the plateaux, dendritic and radial patterns characterize the eroding peatlands and hinder stock control; these areas are of negligible agricultural value.

THE CLASSIFICATION

The classification comprises three main categories, the class, the division and the unit, of which only the first two are utilized on the 1:250 000 map presented with this report. Land placed in any *class* or in any *division* has a similar *overall degree* of limitation; within any class or division there are therefore different management requirements. Comments on the principal *types* of limitation and the management problems which occur will be found in the descriptions of the classes and divisions.

Land in Classes 1 to 4 is suited to arable use and that in Classes 5–7 unsuited to arable use. There are no divisions within Class 1, 2 and 7, two divisions in each of

Classes 3 and 4, and three divisions in Classes 5 and 6. A full description of the classification system and national guidelines is available as a Soil Survey monograph (Bibby *et al.*, 1982). The following is a condensed description of the classes and divisions:

Land suited to arable cropping

Class 1   Land capable of producing a very wide range of crops
Cropping is highly flexible and includes the more exacting crops such as winter-harvested vegetables. The levels of yield are consistently high.

Class 2   Land capable of producing a wide range of crops
Cropping is very flexible and a wide range of crops may be grown but difficulties with winter vegetables may be encountered in some years. The level of yield is high but less consistently obtained than in Class 1.

Class 3   Land capable of producing a moderate range of crops
Division 1   The land is capable of producing consistently high yields of a narrow range of crops (cereals and grass) or moderate yields of a wider range (potatoes, field beans and other vegetables and root crops). Grass leys of short duration are common.
Division 2   The land is capable of average production but high yields of grass, barley and oats are often obtained. Grass leys are common and longer than in division 1.

Class 4   Land capable of producing a narrow range of crops
Division 1   Long ley grassland is common but the land is capable of producing forage crops and cereals for stock.
Division 2   The land is primarily grassland with some limited potential for other crops.

Land suited only to improved grassland and rough grazings

Class 5   Land capable of use as improved grassland
Division 1   Land well suited to reclamation and to use as improved grassland.
Division 2   Land moderately suited to reclamation and to use as improved grassland.
Division 3   Land marginally suited to reclamation and to use as improved grassland.

Class 6   Land capable only of use as rough grazing
Division 1   Land with high grazing value.
Division 2   Land with moderate grazing value.
Division 3   Land with low grazing value.

Class 7   Land of very limited agricultural value.

The following assumptions are made when using the classification:

1   The classification is designed to assess the value of land for agriculture.
2   Land is classified according to the degree to which its physical characteristics affect the flexibility of cropping and its ability to produce certain crops consistently.

170

3   The classification does not group land according to its most profitable use.
4   The standard of management is defined as the level of input and intensity of soil, crop and grassland management applied successfully by the reasonable and practical farmer within the relevant sector of the farming industry. Such management will maintain or improve the land resource.
5   Land with limitations which may be removed or reduced at economic cost by the farmer or his contractors is classed on the severity of the remaining limitations.
6   Land with severe limitations is classified accordingly except where there is clear evidence that a major improvement project (e.g. arterial drainage) will be completed within the next 10 years. In such cases the land is classed as if the improvements had occurred.
7   Location, farm structure, standard of fixed equipment and access to markets do not influence the grading. They may, however, affect land use decisions.
8   The interpretations are expressions of current knowledge and revision may be necessary with new experience or technological innovations.

THE CLASSES AND DIVISIONS IN EASTERN SCOTLAND

The areas of each land capability class and division are shown in Table B and expressed as a histogram in Fig. 11.

Class 1

Covering only 17 square kilometres, Class 1 land is restricted to the sheltered lowlands around Carnoustie where only very minor climatic restraints occur. It totals almost half of the recorded area of this class in Scotland, but less than 0.1 per cent of the total area of Eastern Scotland.

Situated on the 15- and 30-metre raised beaches and the adjacent lowlands, the area has a growing season of about 240 days and receives between 700 and 800 millimetres annual rainfall. Average monthly mean temperatures are relatively high at around 8.4°C, with annual accumulated temperatures above 5.6°C totalling around 2560.

The soils are formed in parent materials derived from Lower Old Red Sandstone sediments. They comprise mainly the freely drained humus-iron podzols of the Panbride and Forfar Associations; also found are imperfectly drained humus-iron podzols and imperfectly drained brown forest soils of the Forfar and Balrownie Associations respectively. Although the subsoils of the Panbride Association are sands and gravels and contain usually less than 10 per cent combined silt and clay, productivity results from artificially deepened topsoils; these have been created often by the addition of seaweed. They are sandy loams more than 50 centimetres thick and offset the limited moisture-and nutrient-holding capacities of the coarse-textured subsoils. Thick topsoils are common also in the soils of the Forfar and Balrownie Associations where the underlying loam to sandy clay loam till facilitates nutrient storage and is an insurance against drought in a very dry season. In these widespread anthropogenic topsoils, stone content is unusually low. The fine crumb structure readily produces excellent seed-beds and permits maximum root development.

Class 2

Within Eastern Scotland, Class 2 land occupies approximately 1100 square kilometres (4.5 per cent of the land area), almost 64 per cent of the the total Class

SUBREGION	LCA CLASSES and DIVISIONS														subregion areas (sq. km)
	1	2	3.1	3.2	4.1	4.2	5.1	5.2	5.3	6.1	6.2	6.3	7	BUA	
MORAY FIRTH LOWLANDS															1,510
SKENE LOWLANDS															1,075
BUCHAN PLATFORM															1,210
UPPER BUCHAN PLATFORM															1,490
FORTH LOWLANDS															695
FIFE LOWLANDS AND UPLANDS															1,650
STRATHMORE AND SIDLAWS															2,960
MORAY FIRTH FOOTHILLS															1,370
GRAMPIAN FOOTHILLS & UPLANDS															4,290
CAIRNGORM MOUNTAINS															1,945
MONADHLIATH MOUNTAINS															2,475
SOUTH-WEST GRAMPIAN HIGHLANDS															3,055
NORTHERN HIGHLANDS															820
LCA total areas (sq. km)	15	1,100	2,770	4,310	1,135	1,580	450	2,155	1,165	270	1,870	5,880	1,640	205	24,545

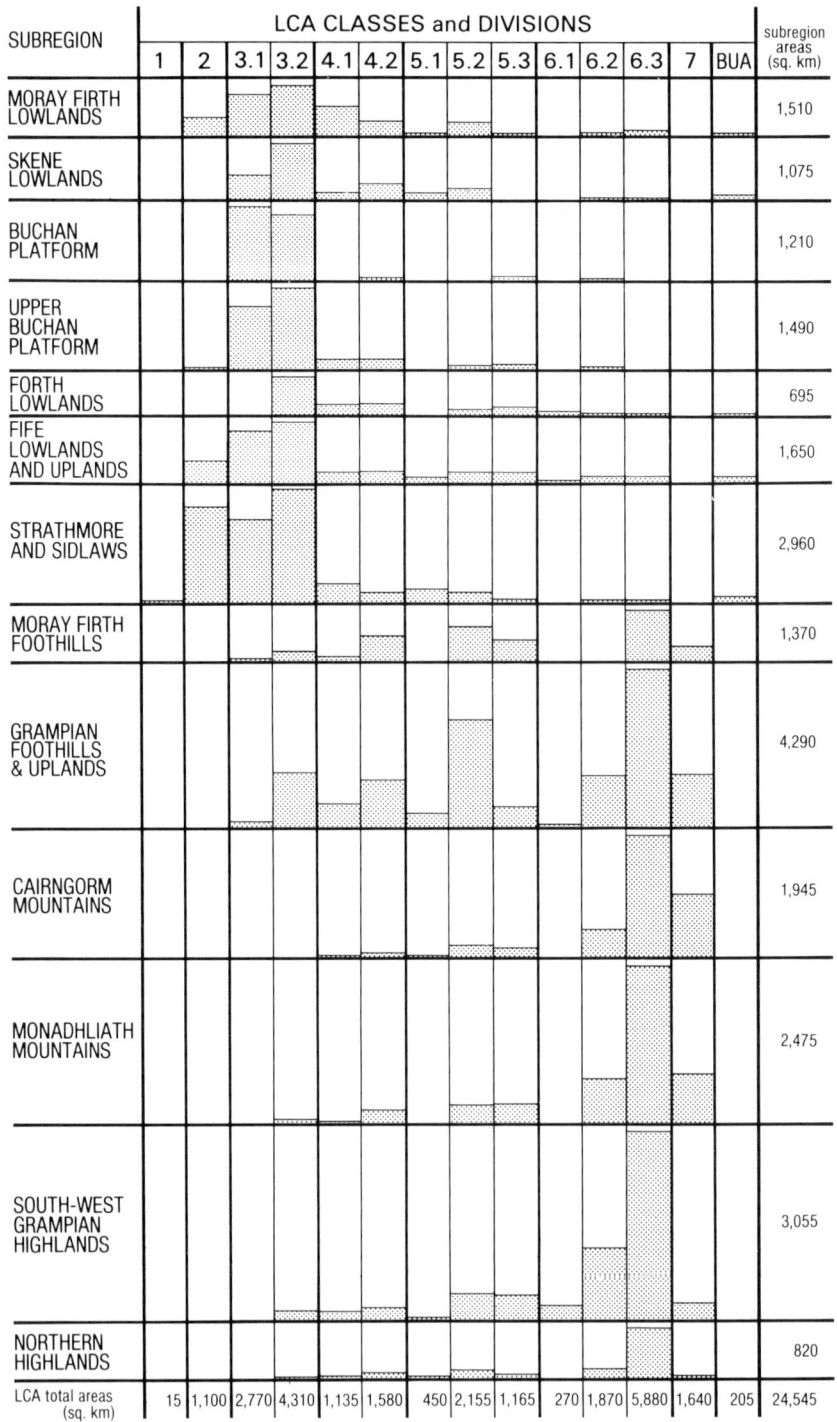

Figure 11. *Histogram of land capability for agriculture map units in the physiographic subregions of Eastern Scotland; subregions are grouped in altitudinal zones. Approximate scale: 1mm = 60sq. km.*

2 area in Scotland. Apart from minor areas in the Garioch district of the North-East Lowlands, it is virtually confined to below 100 metres in the Moray Firth Lowlands and the Central Lowlands. Subject only to minor climatic restraints, nearly all the Class 2 land receives less than 800 millimetres rainfall: minor areas in central Fife and in the north-western part of the Howe of the Mearns receive more, at around 900 millimetres. The coastal zone of the Central Lowlands south from Arbroath, however, has more bright sunshine hours compared with the Moray Firth Lowlands where the marginally less favourable climate is commensurate with the increased northerly latitude.

Major areas of Class 2 land in the Central Lowlands occur in the vale of Strathmore, the Howe of the Mearns, the lower catchments of the Rivers North Esk, South Esk and Lunan, and along the coastal belt between Dundee and Lunan Bay; smaller areas are located to the east of Dundee and in Fife. Altogether they cover some 925 square kilometres.

North of the Tay estuary, the Class 2 land is found mainly on the loam to clay loam tills of the Balrownie and Laurencekirk Associations, the water-worked till of the Forfar Association and the adjoining alluvium. South of the estuary, the areas are restricted, in general, to the clay loam till of the Rowanhill Association on the coastal plain around Kirkcaldy, and to the sands and gravels of the Eckford Association in the Howe of Fife. Small areas are located on the raised beach sands and gravels of the Panbride Association near Carnoustie and St Andrews, and in those of the Dreghorn Association between Largo Bay and Fife Ness. Other small areas occur on the silts, sands and gravels of the Carpow Association along the Tay estuary and on the terraces of the Rivers Earn and Eden. The soils are dominantly freely and imperfectly drained brown forest soils and podzols with the imperfectly drained brown forest soils being the most common. All the tills have medium or moderately fine textures with a characteristic angular blocky structure in the subsoils; they have good moisture- and nutrient-retention capacities. Often the upper horizons consist of water-worked sandy loams and loams up to 45 centimetres thick, their friable consistency facilitating the production of a good tilth and encouraging root development; in the Forfar Association such upper horizons exceed 60 centimetres. On the raised beaches, deep sandy loam topsoils are widespread, thus counteracting the poor water and nutrient capacities of the subsoils which have very low combined silt and clay contents, usually less than 10 per cent.

In the Moray Firth Lowlands, which are also underlain by Old Red Sandstone sediments, Class 2 land covers approximately 145 square kilometres. The individual areas compared with those in the Central Lowlands are smaller but they occupy similar situations. They occur mainly on moderately coarse-textured raised beach and fluvioglacial sands and gravels of the Boyndie and Corby Associations, on lacustrine, estuarine and riverine alluvium, and on the medium-textured tills of the Cromarty and Kindeace Associations; the latter are comparable with the Balrownie and Forfar Associations respectively. Unlike those of the Central Lowlands, the non-alluvial soils are nearly all freely drained iron and humus-iron podzols. Exceptions include the imperfectly drained podzols of the Cromarty Association, which exists only in the Black Isle and around Nigg in Easter Ross, small areas of brown forest soils of the Bracmore Association near Conon Bridge, and the Carden Association near Elgin. The moderately coarse-textured soils are all characterized by deep, friable, man-made topsoils which are usually stone-free and more than 50 centimetres thick. Freely and imperfectly drained riverine alluvial soils are also characterized by similar, but entirely natural, thick fine sandy loam topsoils; these often overlie gravel as in the Nairn and Findhorn valleys. All such deep topsoils counterbalance the inherent limited nutrient- and water-holding capacities of the coarse-textured subsoils. In addition, seed-bed preparation is facilitated by their friable consistency and their highly permeable nature permits access for most of the year.

Class 2 land developed on tills is restricted largely to the Cromarty Association. Frequently, the plough layer has incorporated a coarse-textured morainic material to form a sandy loam topsoil; below 45 centimetres an indurated B horizon merges gradually with the sandy loam to sandy clay loam parent material. The impedance to rapid percolation caused by induration is probably beneficial for these relatively shallow upper horizons in this area of low rainfall. Where these tills occur near ancient settlements, for example, near Auldearn, deep topsoils exceeding 60 centimetres are common.

In sharp contrast to the above, most of the Class 2 land developed in estuarine and lacustrine deposits is poorly drained. Around Beauly, the soils are mainly medium-textured soils of the Nigg Association whereas around Spynie they are fine textured and belong mostly to the Duffus Association; associated with both areas are undifferentiated, poorly drained riverine alluvial soils, usually loamy. Largely because of the low rainfall, around 750 to 800 millimetres, these sites have responded well to trunk drainage schemes. Although the high silt and clay contents in the topsoils would, under a wetter climate, normally induce problems associated with capping and a poor tilth, the effects of drainage and a beneficial climate, coupled with high nutrient reserves and capacity, have resulted in highly productive soils.

Regardless of the parent material, the Class 2 land is devoted primarily to arable and mixed farming with similar ranges and yields of crops in both the Central and Moray Firth Lowlands. Although a six- or seven-course rotation is still widespread, continuous cereals and long-term requirements for soft fruit, 4 years for strawberries and 11 years for raspberries, provide major regional variations in crop patterns. Thus almost 90 per cent of the national crop of raspberries (around 15 000 tons) is grown in Perth, Angus and Fife where it is largely processed for canning and jam-making; soils range from sandy clay loam tills to the sands and gravels of the coastal belt. The associated industrial base has also prompted the development of the local but specialist vegetable crops such as peas for canning and freezing. By contrast, the combined total for soft fruit and vegetables in the Moray Firth Lowlands is only around 200 hectares of which half is soft fruit; much of this acreage has been a response to the recent development of cold storage and freezing plants in and around Inverness. The dominant crops throughout Class 2 land, however, are cereals and grass with barley and wheat having expanded at the expense of oats; in the Moray Firth Lowlands, Angus, Perthshire and Fife a substantial amount of barley is contracted for malting. In addition, the acreage of winter-sown barley is rising steadily with increasing use being made of the late-summer rainfall peaks to cultivate oil seed rape. Grass production is geared to silage although hay-making with cylindrical and oblong bales is still popular. Turnips have declined substantially because of improved grass utilization though the use of precision seeders and mechanical harvesting on the gentle, relatively stone-free slopes has possibly halted the decline. Potatoes are grown for both seed and ware with the production of virus-free seed potatoes being of paramount importance in Angus and Perthshire and constituting most of the crop. The trade is concentrated amongst specialist growers especially where there are deep stone-free topsoils. Irrigation in Strathmore is used mainly on potatoes to increase yield and reduce scab infection at the early stages of growth; in east Fife it is used on a variety of crops especially field vegetables which are grown on contract for a major frozen food processing organization.

Table B Areas of land capability for agriculture map units

CLASS and DIVISION	SHEET 5		SCOTLAND	
	SQ. KM.	% LAND AREA	SQ. KM.	% LAND AREA
1	17	0.1	41	0.1
2	1098	4.5	1723	2.2
3	7076	28.8	11724	15.2
3.1	2768	11.3	4586	5.9
3.2	4308	17.5	7138	9.3
4	2717	11.1	8219	10.7
4.1	1135	4.6	3690	4.8
4.2	1582	6.5	4529	5.9
5	3775	15.4	14270	18.5
5.1	454	1.8	1810	2.4
5.2	2155	8.8	5899	7.6
5.3	1166	4.8	6561	8.5
6	8017	32.6	37329	48.4
6.1	270	1.1	1556	2.0
6.2	1870	7.6	5463	7.1
6.3	5877	23.9	30310	39.3
7	1636	6.7	2548	3.3
BUILT-UP AREAS	208	0.8	1233	1.6
TOTAL	24544		77087	

1 sq. km. = 100 hectares

Areas in this table have been estimated by point-count methods. Care should be exercised in calculations involving units of less than 10 square kilometres. Discussion of method and estimation of error is contained in Handbook 8.

175

Class 3

Class 3 land is capable of producing a moderate range of crops and is the backbone of the arable and mixed farming sector in Scotland, occupying almost 12 000 square kilometres. Some 60 per cent of that total area (7076 square kilometres) occurs in Eastern Scotland where it is found mainly below 300 metres in the Moray Firth and North-East and Central Lowlands. In the Highlands, minor areas are located in the major valleys such as those of the Rivers Spey, Tay and Conon. There are only minor and moderate climate constraints and rainfall is mostly below 900 millimetres though in the western parts of the Upper Buchan Platform it reaches 1000 millimetres, and in the upper Forth valley and upper Strathearn it may exceed 1200 millimetres. In these western areas, stock-rearing is dominant. Although limitations regarding choice of crop and yield are more restrictive in Class 3 than in the previous classes they are still only moderate in degree and a wide range of soil types is included. Their properties range, for example, from the clay textures and massive or strong prismatic structures of the Carse clays to the coarse textures and single-grain structures of many arenaceous raised beach soils, from a restricted rooting volume, often due to induration, to the unrestricted rooting volume of the fluvioglacial sands of Moray and Nairn and from the freely drained, strongly sloping ground of the Grampian Foothills to the low-lying sumps with poor drainage so common in the Buchan Platform. Despite such variations, generalizations may be made. Thus, arable farming is concentrated in the gentle lowlands below 200 metres with grass conservation and stock-rearing common above that altitude. Grass is also increasingly important in the wetter west especially in the Carse of Stirling where Timothy hay and seed production forms part of a specialized rotation.

Division 1 Class 3.1 land is widespread in Eastern Scotland, covering 2768 square kilometres; this is more than 60 per cent of the total area of Class 3.1 land in Scotland.

In the Central Lowlands, where it occupies about 1080 square kilometres, division 1 land occurs mainly on the undulating plains and is dominated by imperfectly drained, moderately fine-textured brown forest soils of the Balrownie and Rowanhill Associations (Plate 17). Smaller areas which are associated with higher sloping ground largely comprise moderately coarse- to moderately fine-textured soils of the Darleith, Mountboy and Sourhope Associations; these soils include both free and imperfectly drained brown forest soils. Fine-textured soils, mainly of the Stirling Association, are confined to the Carses of Gowrie and Earn whereas coarse-textured soils are scattered throughout the subregion. Much of the division is closely related to Class 2 land, often occupying slightly less favourable sites where increased altitude, rainfall and exposure reduce crop selection and performance.

Of the moderately fine-textured soils, those of the Balrownie Association form a broad swathe from Dundee to Arbroath but have a patchwork distribution in northern Strathmore; other areas occur in Strathearn and near Luncarty. In the Rowanhill Association, which is restricted to Fife, division 1 land is found mainly in the south-east peninsula. Although subject to relatively low rainfalls of 700 to 800 millimetres, the good water-holding capacity of all these sandy clay loam and clay loam subsoils counteracts droughtiness. Some areas have water-sorted, upper horizons whose coarser texture improves accessibility and facilitates tilth production.

The higher ground associated with division 1 land is located on the flanks of

Plate 17. *Undulating lowlands south-east of St Andrews. There is a mixed pattern of agriculture with the emphasis on arable farming. Brown forest soils with gleying, map units 444 and 445 (Rowanhill Association, Classes 3 and 4) are dominant; there are minor areas of finer-textured, noncalcareous and humic gleys, map unit 445. In the distance is Largo Law and soils of the Darleith Association.* Cambridge University Collection: copyright reserved.

the Ochil Hills and the Sidlaw Hills. There, the imperfectly drained brown forest soils, with sandy clay loam subsoils and belonging to the Sourhope and Mountboy Associations, occupy the lower gentle slopes. Further upslope, the soils are freely drained brown forest soils of the same associations with stony sandy loam textures; in the Ochil Hills similar soils of the Darleith Association are included. Moderate limitations in these higher soils are stoniness, shallowness due to induration, and complex slope patterns.

Fine-textured soils are restricted to the Carse of Gowrie and the adjacent lower reaches of the River Earn where they comprise poorly drained clays and silty clays with well-developed coarse prismatic structures. Wetness and structure limitations lead to difficulties in seed-bed preparation and in root crop germination due to capping, and a high standard of management is necessary to achieve the high potential productivity Gauld *et al.*, (1983).

The only major area of coarse-textured soils is a complex of podzolic sands and gravels of the Corby and Boyndie Associations and alluvial soils at the junction of the Rivers Isla and Tay. Freely drained brown forest soils of the Carpow Association occupy lower areas on the sand and gravel terraces of the River Earn. Minor areas of Class 3.1 land occur also on similar brown forest soils of the Darvel

177

and Eckford Associations in Fife. All these coarse-textured soils are easily worked and accessible throughout the year; the main limitations are texture, drought and stoniness. Unlike the granite and schist-derived sands and gravels of the northern areas of Eastern Scotland, soils of the Corby and Boyndie Associations in the Central Lowlands are less prone to trace-element deficiencies because of the argillaceous content of the parent material.

Much of the lower land in division 1 in the Central Lowlands is devoted to arable cropping, mainly barley production, with some mixed farming and localized areas of soft fruit; the fruit is concentrated in the southern sectors of Strathmore. Indeed the Carse of Gowrie was once renowned for top-fruit orchards, now much depleted because of adverse economic factors. Occasionally, where the tills are bordering upon the fine textures or where altitude and rainfall increase, a 7- or 8-course rotation is still practised. Thus, as the ground rises in the western and northern areas, the land becomes marginal for the division and the climatic limitations increasingly promote the fattening of livestock within the system. This is also the case around the footslopes of the Sidlaw Hills and Ochil Hills where stoniness, coarser textures and increasing complex slopes encourage the development of grass in the rotation.

Within the Moray Firth Lowlands, Class 3.1 land covers 330 square kilometres but, in sharp contrast with the North-East and Central Lowlands, most of the soils are freely drained podzols with coarse and moderately coarse textures; alluvial soils are locally common, ranging from sands to silts and varying in drainage from free to poor. The low rainfall of 600 to 800 millimetres often induces droughtiness in the freely drained, coarse-textured soils. East of Inverness, the division is found mainly on the coastal lowlands below 76 metres on fluvioglacial sands and gravels of the Corby and Boyndie Associations and on sandy, occasionally loamy, alluvium associated with the lower reaches of the Rivers Nairn, Findhorn, Lossie and Spey. Minor areas occur on fluvioglacial silts of the Polfaden Association and on sandy loam tills of the Elgin and Kindeace Associations. North of the Moray Firth, the division occupies much of the lower slopes of the Black Isle where the gradient is often near the upper limit of 7 degrees. The soils are, respectively, stony sandy loam and loam tills of the Millbuie and Cromarty Associations. Shallowness, caused by induration, and stoniness are the main limitations but the soils are easily worked throughout the year. From Nigg to Inverness, much of the division occurs on the poorly drained sandy and silty low raised beaches of the Nigg Association and the related, and undifferentiated alluvium of the lowest reaches of the Rivers Beauly, Conon and Peffer where wetness and poor structure are the main limitations but where excellent yields of cereals are obtained; the adjacent higher raised beaches are formed in freely drained sands and gravels of the Corby Association. Minor areas are located on the stony, sandy loams of the Kessock, Orton and Sabhail Associations on the nearby gentle slopes where stoniness, induration and moderately coarse textures are limiting. Fine-textured brown forest soils of the Braemore Association occur between Contin and Beauly; together with those of the Carden Association near Elgin, they form the only significant areas of such inherently fertile soils in the Moray Firth Lowlands. Capping, which is a primary limitation in these soils and in much of the poorly drained estuarine alluvium, results from the high silt contents.

Across the Moray Firth Lowlands, the emphasis is on arable farming and stock fattening using a 6-course rotation but with an increasing bias towards continuous barley production. Thus, the development of major malting plants coupled with suitable varieties have led to a substantial rise in cereals from 50 000 to 60 000 hectares in the Moray Firth area between 1970 and 1980. Barley has become dominant at the expense of oats and turnips and has risen from 24 000 to 49 000 hectares in the same period; similar trends have occurred in the other lowland subregions. Seed potatoes are of considerable local importance on the sandy raised beaches. The only other vegetable crop of importance is carrots with the soils of the Boyndie Association being particularly suitable; Morayshire accounts for almost 12 per cent of the Scottish crop.

In the North-East Lowlands, Class 3.1 land is mainly below 200 metres within a rainfall of 800 to 900 millimetres. It occurs mostly on freely and imperfectly drained brown forest soils with medium to fine textures; all these soils are of inherent superior fertility, belonging to the Insch, Tarves, Peterhead and Tipperty Associations and to a lesser degree, the Foudland Association. Minor areas occur on moderately coarse-textured soils, usually podzols.

On the Buchan Platform, Class 3.1 land covers 580 square kilometres. In the Insch valley, the soils are freely drained brown forest soils, some with plaggen topsoils, and belong to the Insch Association; they are highly productive and easily worked. Over much of the undulating ground between the Rivers Don and South Ugie, where the Tarves Association is paramount, the soils are mainly freely drained stony sandy loams but also include imperfectly drained brown forest soils with a sandy clay loam subsoil which becomes massive with depth. Induration is widespread and restricts rooting volume. On the coastal belt north of the Ythan estuary, are imperfectly drained lacustrine clay loams and clays of the Tipperty Association; to the south, are similar clay loams of the related Peterhead Association. Despite the low rainfall, these fine- and moderately fine-textured soils require careful management to create seed-beds and autumn ploughing is essential to promote a tilth by frost action. Surface caking is a major limitation affecting the germination of root crops in a wet spring.

In the Upper Buchan Platform, the Class 3.1 land occupies some 500 square kilometres, mainly in a belt of friable sandy loams of the Foudland and Ordley Associations which stretches from the Howe of Auchterless almost to the coast. These soils have a low stone content and are readily worked, with those of the Ordley Association comprising some of the most productive land in the North-East Lowlands; arable cropping and stock-fattening are the main enterprises.

In the Skene Lowlands, the division 1 land covers about 195 square kilometres. Excepting the dominant brown forest soils of the Tarves Association in the Alford basin and along the northern fringe, the area consists largely of freely drained humus-iron podzols of the Countesswells Association with stony, sandy loam textures; stoniness, and shallowness caused by intense induration are characteristic limitations and stock-fattening the dominant activity.

As in the Moray Firth, a 6-course rotation is traditional in the North-East Lowlands. The region, however, is more exposed especially around the coastal zones and is subject to moderate climatic constraints as compared with the very minor and minor climatic constraints which operate over much of the Moray Firth Lowlands and the Central Lowlands. These climatic limitations coupled with the land tenure based on family farms have led to the development of a flexible agricultural system based on livestock especially the breeding, rearing and fattening of beef cattle. On the northern part of the Buchan Platform and the eastern areas of the Insch valley where the freely drained brown forest soils permit a greater flexibility in feed production, the fattening of cattle is the main enterprise though fat lamb production and breeding of lowland sheep are often integral parts of the system. Around the Don valley and the southern section of the Tarves Association, the fine textures of the imperfectly drained brown forest soils are eminently suited to grass production and dairying is dominant; the milk-processing industry in Aberdeen and the large consumer population ensure a relatively steady market. On the coastal belt, the imperfectly drained clays and clay loams of the Peterhead and Tipperty Associations form some of the most naturally fertile land in the north-east. Though demanding a high level of management especially in cultivation, the area is capable of producing high yields of grasses and cereals with root crops less successful. Haar, low stratus cloud sweeping in from the sea, and exposure affecting harvest are limiting factors additional to the principal ones of texture and structure. Arable cropping and the fattening of cattle are the main enterprises.

Division 2 This land covers 4308 square kilometres in Eastern Scotland (17.5 per cent of the land area) and forms 60 per cent of the Class 3.2 land in Scotland. It occurs throughout the lowland subregions, the Moray Firth Foothills and the Banffshire section of the North-East Grampian Foothills; it also includes minor areas in the valleys of the Rivers Findhorn, Spey, Dee and Tay. In general, it is linked with the freely drained coarse- and moderately coarse-textured podzolic soils, especially on the higher sloping ground, and with poorly drained moderately fine- and fine-textured soils in low-lying areas; in addition it dominates the western Forth valley where the rainfall exceeds 1000 millimetres.

Covering almost 1590 square kilometres in the North-East Lowlands, the Class 3.2 land is associated in the extreme north-east with free-draining stony, sandy loam podzols of the Corby and Strichen Associations. On the Upper Buchan Platform, it occurs on similar moderately coarse-textured soils of the Foudland Association and, in the Skene Lowlands, on those of the Countesswells Association. The till-derived soils are usually shallow, 1.25 metre or less thick, and induration is widespread, being weak in the slate-derived soils and intense in the granite soils; the latter have the additional limitation of extreme boulderiness. In some of the more moundy areas of the Corby Association, cultivation leads to the erosion of the topsoil from the local summits. All these soils are easily worked and accessible throughout the year.

Over much of the Buchan Platform and Insch valley, the Class 3.2 land is located on poorly drained noncalcareous gleys of the Peterhead, Tarves and Insch Associations. Situated in level sites or shallow depressions, the clay loam and loam topsoils overlie massive clay and clay loam subsoils. Poaching, trafficability problems in winter and difficulties in seed-bed preparation are major limitations.

In the Central Lowlands, the division totals approximately 1670 square kilometres and is widespread throughout the subregion. It is common on the imperfectly drained and moderately fine-textured brown forest soils of the Balrownie Association from the Forth valley to the North Esk and on the Stonehaven Association between Montrose and Inverbervie. Because of the poorly structured sandy clay loam subsoils, drainage is slow and can restrict access and hinder cultivation especially in the wetter south-west; in the Stonehaven Association, cobbles are an added hindrance. The Gourdie Association, with similar textures, occupies the foothills along the Highland Boundary Fault from Kirriemuir to Glen Artney, with gradient on the upper slopes limiting the range of crops. Within Fife, Class 3.2 land occurs on slightly finer-textured soils which belong to the Rowanhill Association; there the clay loams extend from the east Fife plateau across the coastal lowlands to Alloa. In northern Strathmore and parts of southern Fife, the upper horizons of such fine-textured subsoils are overlain occasionally by water-worked sandy loam horizons which facilitate a more rapid drainage and hence a greater cropping flexibility.

Moderately coarse-textured soils, mainly podzols of the Forfar and Strathfinella Associations, are found, respectively, throughout Strathmore and on the footslopes along the Highland Boundary Fault north of Glen Clova. The common limitations are texture, stoniness and induration, though the additional restriction of slope in the Strathfinella Association limits cropping to rotations based on long ley grassland. A similar situation is found on the moderately coarse-textured soils of the Sourhope Association on the slopes of the Ochil Hills. Coarse-textured soils include the brown forest soils and minor humus-iron podzols of the Gleneagles Association mainly in Strathallan and those of the Doune Association in the upper Teith valley. Podzols of the Corby/Boyndie Associations are scattered across the subregion north of Perth. Pattern, often involving wet hollows within hummocky terrain, stoniness, shallow topsoils and coarse textures are the main limitations of these fluvioglacial sands and gravels and partially restrict cereal cropping; stock-rearing and fattening are the main enterprises.

Around the Moray Firth, Class 3.2 land is related principally to freely drained and podzolic coarse- and moderately coarse-textured soils; in the Moray Firth Lowlands it accounts for 390 square kilometres. From Portgordon to Inverness, it is associated with the sands and gravels of the Corby and Boyndie Associations and, to a lesser extent, with the partially sorted sands and gravels of the Ardvanie, Brightmony and Dulsie Associations; texture, stoniness and droughtiness affect all these fluvioglacial soils, with moundy topography often an added limitation; the crests of the mounds are subject to erosion by cultivation operations especially by ploughing. In the Black Isle and on the foothills surrounding and occasionally projecting through the lowlands, are moderately coarse-textured tills of the Kindeace, Elgin, Orton and Sabhail Associations; shallow topsoils, caused by induration, and stoniness are the main limitations. Eastwards, into the Banffshire foothills and uplands, podzolic, stony sandy loam tills of the Strichen, Durnhill and Foudland Associations comprise the Class 3.2 land on the slopes whereas in the depressions between Huntly and Portsoy are noncalcareous gleys of the Strichen and Tarves Associations with clay loam subsoils. In these latter areas, wetness coupled with the slightly higher rainfall of 900 millimetres and exposure to the cold north-easterly wind create cultivation and harvesting problems.

Class 3.2 land is capable of growing a moderate range of crops and is generally associated with relatively small family-farms where social factors have led historically to flexibility based on mixed farming. In many areas, however, the limitations of climate and soil, especially texture and rainfall, dictate an emphasis on grass and winter feed. On the lower ground, the more sheltered environment allows a greater variety of cropping especially where moderately fine-textured soils are subject to only the lower range of rainfall, for example, much of the Balrownie Association in Strathmore and the Rowanhill Association in eastern Fife. There, arable cropping, with barley for malting and stock feed as a principal crop, is integrated with cattle-fattening, sheep-rearing and the production of fat lambs. On the slopes along the edge of the Grampian Highlands and on the isolated hill ranges such as the Ochil Hills, the emphasis gradually changes with increasing slope and altitude from fattening to rearing of stock. Another system gradient develops westwards from west Fife to the upper Forth valley where as the rainfall rapidly increases, grass production and conservation become more important. Although arable crop yields are high on

181

these fine-textured soils, especially on the Carse of Stirling, the problems in cultivation and harvesting associated with the wetter climate are correspondingly greater and demand a high standard of management. Dairying is important in west Fife, mainly on the Rowanhill Association, where the fine-textured and moderately fine-textured soils are particularly suited to grassland management and the large industrial population stimulates the demand for milk products; the situation is paralleled on the Tarves Association north of Aberdeen.

Class 4

Class 4 land is capable of producing a narrow range of crops based primarily on grassland but with a limited potential for forage and cereal crops which is generally fully exploited in the drier areas. The farming is geared mainly to stock-rearing and breeding.

It occupies some 2717 square kilometres of Eastern Scotland (11 per cent of the land area) and is related largely to the peripheral slopes of the lowlands bordering the mountains and the isolated hills and ranges within these plains. Some 60 per cent of the class occurs in the lowland subregions (Fig. 2) where it forms about 15 per cent of their land area; the remainder is found in the foothills of the Grampian Highlands and the Northern Highlands with minor areas occurring along the valley floors and adjacent slopes of the major rivers (Plate 18); it rarely occurs above 350 metres and the rainfall is usually less than 1000 millimetres. The class is associated mainly with increasingly adverse climatic restrictions correlated with altitude and exposure, except in the upper Forth valley where rainfall between 1200 and 1600 millimetres severely restricts the flexibility and selection of crop. With a growing season progressively reduced to as little as 190 days in such places as the upper Dee valley and Glenlivet at around 300 metres, as opposed to over 240 days at the coast, there are obvious potential hazards in the timing of seed-bed preparation and harvesting. Yields, especially of roots and cereals, are very variable; in the wetter areas, soils are also fine textured and additional problems are likely in the conservation and utilization of grassland.

Division 1   Class 4.1 land in the North-East Lowlands covers slightly less than 100 square kilometres in both the Upper Buchan Platform and the Skene Lowlands but is of negligible extent in the Buchan Platform. In the Upper Platform, it is closely associated with shallow, stony sandy loam drifts of the Foudland Association on relatively exposed and isolated summits; the largest area occurs in the Glens of Foudland. Within the Skene Lowlands, Class 4.1 land occurs commonly on undulating slopes and on the flanks of granite hills such as the Hill of Fare and Bennachie. In Deeside, it occurs also on the terraced and moundy sands and gravels of the Corby/Boyndie/Dinnet Associations. Most of the drift soils are till-derived, freely drained humus-iron podzols, the limitations of stoniness, and shallowness resulting from induration being widespread; the fluvioglacial soils are also susceptible to drought and trace element deficiencies. Though grassland is paramount, barley and oats, together with root crops, are widely grown for stock-feed in the dominant enterprises of beef cattle and sheep-rearing and breeding.

In the Central Lowlands, Class 4.1 land is virtually absent from the Strathmore plain though the lower foothills of the Grampian Highlands form a prominent belt of this division along its western rim between Dunkeld and Glen Clova. There, freely drained humus-iron podzols and brown forest soils of the Strichen and Gourdie Associations are subject to limitations of shallowness due to rock or

Plate 18. *The upper Don valley west of Strathdon, showing the typical land uses of agriculture, forestry and grouse-moor. The flat arable fields in the foreground are on the brown forest and alluvial soils of map unit 422 (Nochty Association), Class 4.1. In the middle distance, the soils on the slopes are brown forest soils and humus-iron podzols of map units 316 and 318 (Insch Association), 520 (Tarves Association) and 243 (Foudland Association) — mainly Classes 4 and 5. The tors of Ben Avon in the Cairngorm Mountains are visible on the skyline.* Photograph by kind permission of Aberdeen Journals Ltd.

induration, and stoniness; occasionally pattern caused by hummocky moraines is associated with the Strichen Association. Because of the relatively low rainfall of around 900 millimetres and the free-draining nature of the dominantly moderately coarse-textured soils, a fairly flexible 7- or 8-year rotation is possible and is related to the rearing and breeding of beef cattle and sheep. On the low ground, minor areas occur where shallow, stony, sandy loam drifts of the Balrownie and Mountboy Associations overlie rock; around Edzell, an outwash plain of the River North Esk comprises podzolic soils on sands and gravels of the Corby and Boyndie Associations. Stoniness, texture and droughtiness are the principal limitations of these low-ground areas.

In north and central Fife, Class 4.1 land is associated with the lower flanks of the Ochil Hills and Lomond Hills where freely drained brown forest soils are formed mainly in the stony, sandy loam drifts of the Sourhope and Darleith Associations respectively. Minor areas include coarse-textured soils of the Darvel Association at the base of the steep southern slopes around Dollar and the sands of the Eckford Association in the outwash plain of the Howe of Fife. Slope, pattern and texture are all major limitations affecting these soils, with droughtiness an additional factor in the Howe of Fife where the rainfall is less than 800 millimetres. Stock-rearing of beef cattle and sheep is the dominant activity with stock-fattening locally important. Between Clackmannan and Cowdenbeath, however, the combination of fine-textured soils of the Giffnock

Association with a rainfall of around 1000 millimetres determines the dominance of long ley grassland. Dairying and finishing store lambs are the main farming enterprises. Rainfall increasing westwards in the upper Forth valley and in Strathallan on the dominantly imperfectly drained, moderately fine-textured soils of the Balrownie Association, also dictates the increasing use of long-term grassland. On the finer-textured soils of the Stirling Association near Buchlyvie, the rotation may include grass, especially Timothy, for 8 to 15 years. Only on the freely drained mounds around the Lake of Monteith on the coarse-textured soils of the Callander Association is a more flexible rotation possible. Apart from the production of Timothy seed on the clay soils, the Forth Lowlands are devoted mainly to stock-rearing and breeding and some dairying.

In the southern sector of the North-East Grampian Foothills, apart from Class 4.1 land along the junction with the Central Lowlands, the division is restricted largely to the floors of narrow glens such as Clova, Garry and Isla where podzolic soils on moundy and terraced sands and gravels of the Corby Association permit a moderately flexible rotation of grass, cereals and root crops; limitations include coarse texture, stoniness and liability to flash-flooding. Across the wider valley of the mid Tay and Tummel, however, Class 4.1 land also occupies the lower slopes and hummocky moraine above the valley floor; the soils are moderately coarse-textured brown forest soils and humus-iron podzols of the Strichen Association. Excepting the hummocky moraine, these patterns are repeated throughout the foothills, mainly involving moderately coarse-textured and freely drained podzolic soils of the Arkaig, Foudland and Strichen Associations although the more base-rich soils of the Insch and Tarves Associations are associated with the upper catchment of the River Don. Farming is concerned primarily with stock-rearing, with the north-eastern areas renowned for the breeding of Aberdeen Angus cattle.

Around the Moray Firth, Class 4.1 land occupies approximately 400 square kilometres across the coastal lowlands and the adjacent foothills. It occurs mostly below 250 metres within a rainfall of 650 to 1000 millimetres. In common with Class 4.1 land throughout Eastern Scotland, except southern Fife and the Forth Lowlands, most of the soils are moderately coarse textured, and those of the coastal areas are distinctly coarse textured. Between Burghead and the River Spey, Class 4.1 land is found on the freely and excessivly drained sands and gravels of the Corby and Boyndie Associations with limited areas of poorly drained raised beach sands around Lossiemouth. The former areas have thin topsoils coupled with a weak structure and a low organic matter and are highly susceptible to drought, wind erosion and copper and cobalt deficiencies; in some areas slope and pattern are added limitations. 'Scorching', burial of seedlings, or both, sometimes necessitates the re-establishment of crops. The poorly drained areas suffer from a high ground-water table and weak structure; both types of deposit have a poor nutrient-retention capacity. Despite these limitations, cereal and root crops are widely grown with the emphasis on cattle- and sheep-rearing and fattening.

On the slopes fringing the coastal lowlands and on the adjacent foothills from Alness to Buckie, Class 4.1 land is largely correlated with stony, coarse sandy loam, occasionally loam, drifts of various sandstone-derived associations; these include Elgin, Kessock, Kindeace and Sabhail/Mount Eagle Associations and the mixed sandstone and schist drifts of the North Mormond/Orton Associations.

Most of the division comprises freely drained humus-iron podzols which are normally stony and very stony and shallow, the shallowness resulting from strong induration or rock as is the case in the upper slopes of the Black Isle. Slope is frequently near the class limit especially on complex slopes west of Inverness. Only around the slopes of the Hill of Wangie near Elgin does wetness become a moderately severe limitation. There, noncalcareous and humic gleys with loam textures of the Arkaig Association occupy the lower ground below spring lines but have responded to extensive drainage systems. Stock-rearing is still the dominant activity though grass plays a larger role in the rotation because of the greater influence of slope and shallowness limitations.

Division 2   This land is primarily devoted to grassland though there is a limited potential for other crops. Unlike the wetter, western and south-western areas of Scotland, Class 4.2 land in Eastern Scotland rarely receives more than 1000 millimetres rainfall except in the Forth Lowlands which receive up to 1600 millimetres. Cereals and root crops for stock-feed are therefore often risked on coarse and moderately coarse-textured soils, despite the problems in sowing and harvesting and the unpredictable yields. In the higher rainfall areas, where fine- and moderately fine-textured soils dominate, then grass is grown almost exclusively except for a break-crop prior to replacing the sward. Because of the history of land tenure, coupled to the maintenance of the grouse-moors over much of the northern and north-eastern peneplains, there have been few recent attempts at reclamation for cropping as opposed to many for grassland and the reinstatement of abandoned hill farms. It is therefore possible that the guidelines have been interpreted too severely on occasions and that the area of Class 4.2 land has been underestimated.

Class 4.2 land in the Central Lowlands covers approximately 300 square kilometres divided equally amongst its subregions (Fig. 2). In Strathmore, the division occurs in widely separated areas which are subject to the same type of limitations, but of increased degree, as those which qualify the Class 4.1 land. At Rattray, for example, a pitted outwash plain of sand and gravel of the Corby/Boyndie Associations resembles that at the North Esk but the hollows are deeper and wet. A similar complex of sands and gravels of the Gleneagles Association occurs around Braco; in the Howe of Fife, the excessive stoniness and pattern of fluvioglacial deposits of the Eckford Association have effectively downgraded one major area from the surrounding Class 3.2 land developed in gentle mounds of sand. All these deposits are subject to limitations of pattern, stoniness, droughtiness and poor water- and nutrient-retention capacities. In addition, throughout Strathmore, there are a number of local summits where slope and shallowness, due to rock make tillage difficult and where increased exposure restricts yields and timing of operations. For example, on the northern slopes of the Sidlaw Hills, brown forest soils with very stony sandy loam textures and belonging to the Mountboy, Sourhope and Darleith Associations overlie rock at depths between 20 and 40 centimetres. Similar limitations qualify areas of the Strathfinella Association north of Tullo Hill and areas of the Stonehaven Association south-west of Stonehaven. Wetness limitations on any scale in Class 4.2 land in Strathmore are virtually confined to Glen Almond where the poor hydraulic conductivity of the moderately fine-textured gleys of the Balrownie Association, in combination with the increased rainfall, produces problems in seed-bed preparation and liability to poaching.

In Fife, the same kinds of restrictions qualify Class 4.2 land as in Strathmore, though the division is more frequently found linked to areas of Class 4.1 land. For example, within the Ochil Hills and Lomond Hills it largely reflects steeper and more complex slopes and increased stoniness of the brown forest soils of the respective Sourhope and Darleith Associations whereas in the west Fife plateau,

Class 4.2 land reflects the increasing wetness in the moderately fine-textured soils of the Giffnock Association. To the west of Cowdenbeath, the noncalcareous and limited peaty gleys fill small linear depressions, forming a complex pattern with the surrounding imperfectly drained brown forest soils. West of Saline, however, Class 4.2 land comprises uniform areas of noncalcareous gleys. The combination of texture and wetness severely restricts cropping to long ley grassland because of capping, poaching and related problems. Subsidence resulting from coal-mining is an additional hazard.

Class 4.2 land in the Forth Lowlands forms an aureole around the carse clays and is related basically to the increasing rainfall westwards and its interaction with the fine- and moderately fine-textured soils, though moderately coarse-textured soils with pattern limitations are important around Callander. From the Braes of Doune, the imperfectly drained brown forest soils of the Balrownie Association are rapidly replaced westwards by wetter soils whereby, east of Loch Lomond, Class 4.2 land is dominated by poorly drained and very poorly drained gleys. Similar wet soils of the Kippen and Sorn Associations characterize the lower slopes surrounding the northern flanks of the Gargunnock Hills and very poorly drained clays of the Stirling Association abut Flanders Moss. All these soils are limited by wetness and the attendant problems of poaching, capping and timing of operations consequent upon the effects of higher rainfall upon such textures. Increased organic matter in topsoils also accompanies the westward increase in rainfall so that the fields contain a higher percentage of peaty gleys and humic gleys and the bearing strength of the sward is further reduced. Around Callander, these climatic limitations are mitigated by the the better drainage associated with the moundy and moderately coarse-textured soils of the Callander Association but the moderately severe pattern limitations, involving frequent wet hollows, restrict cropping.

Covering some 255 square kilometres in the North-East Lowlands, Class 4.2 land is concentrated in the Skene Lowlands subregion with lesser areas in the Upper Buchan Platform and a minimal area on the Buchan Platform. In the latter, it occurs mostly near Rattray Head and Balmedie; in the stabilized aeolian sands of the Nigg Association they are prone to wind erosion and droughtiness and have poor water- and nutrient-retention capacities. Minor areas are related to gleys of the Tarves and Insch Associations in wet hollows, to relatively exposed rock-controlled hills of the Durnhill and Leslie Associations and to a kettlehole complex of the Corby Association. In the Skene Lowlands, Class 4.2 land is divided broadly into low-lying, wet areas of granite-derived, sandy clay loam tills, often incorporating peat and alluvial soils, and higher relatively exposed slopes and hill tops. The latter are dominant and are common around the Alford and Tarland basins where the soils are humus-iron podzols.

Class 5

Class 5 land is restricted to a grassland economy but forms an integral part in the hill-farming sector, often playing a vital role in the glens of the Grampian Highlands and Northern Highlands where the valley floors and lower slopes are reserved for the production of winter keep (Plate 19).

Although the land is unsuitable for arable cropping, it is capable of reclamation and improvement by mechanized treatments. These range from complete inversion by ploughing with the associated practices of discing, rolling and harrowing to less intensive systems ranging from simple rotavation to discing by heavy-duty discs and scarifying. Ploughing operations themselves vary from the use of single- or multiple-furrow mouldboard ploughs to a double pass involving separate deep tining followed by mouldboard ploughing and a single pass by a tine-mouldboard plough. The least intensive systems include the killing of the natural vegetation by herbicides, commonly paraquat and glyphosate, followed by minimal scarifying, and the spreading of lime and particularly phosphatic fertilizers. In Eastern Scotland, where the soils in most hill land are acidic, podzolized and moderately coarse textured, a standard prescription includes 7.5 tonnes of lime per hectare, 0.4 to 0.6 tonnes of a general phosphate fertilizer and approximately 0.4 tonnes of a general compound fertilizer. On sites where boulders are a problem, windrowing of the stones by a bulldozer followed by heavy-duty discing has proved highly successful. The recent and rapid development of robust and tracked machinery, often initiated by forestry requirements, and its deployment by specialist contractors has encouraged many grassland reclamation schemes by hill farmers whose limited equipment would be incapable of tackling such projects.

Pioneer crops such as rape and stubble turnips are standard options in most areas, though Class 5 land is generally unsuited for an arable rotation. Though grass yields are highly variable according to local conditions and conservation may be possible in particularly favourable areas, difficulties in utilization are common.

Allocation of land to this class simply signifies that it is suitable for improvement and not that it should be improved. Much of Class 5 land in Eastern Scotland is still untouched by reclamation, a reflection of a number of issues such as land tenure, the objectives of both landlords and tenants especially with regard to sport, not least the financial structure and foreseeable profitability of the farm. Of critical importance is the often conflicting interests of agriculture, sport and forestry, with integration a concept often debated but seldom realized. At farm level and outwith these issues, the decision to reclaim is primarily dependent upon the ease of establishment, the maintenance of the improved sward especially related to the persistence of the sown species and the flexibility available in utilization. Within Eastern Scotland, conservation, though practised on some of the lower, non-rocky slopes, is not widespread and most of the improved grassland is used for grazing.

Generally, ease of establishment is related to freely drained gentle slopes with adequate moisture reserves where machinery can be deployed without damage to soil structure and where the tensile strength of the sward resists poaching. Although much of Eastern Scotland below about 450 metres is capable of reclamation, moderately severe and severe climatic constraints, mainly related to increasing altitude, result in an increasing thickness of surface peaty horizons and a rapidly decreasing growing season. Wet soils, especially fine-textured soils with organic topsoils, are poor reclamation subjects due to the low bearing strength of the sward and the tendency to poach because of poor hydraulic conductivity. Poor oxygenation in waterlogged conditions also leads to reduced biotic activity and sharply reduces the persistence of sown species under competition with the natural vegetation.

Division 1 Class 5.1 land is well suited to reclamation and use as improved grassland and occupies some 454 square kilometres (1.9 per cent of the land area) in Eastern Scotland. Naturally of limited occurrence in the lowland regions where soils and climate are mainly conducive to higher classes, the division nevertheless occupies some 260 square kilometres in these areas of which more than half is associated with the gentle and strong slopes of the Sidlaw Hills, Ochil Hills and

Plate 19. *A section across the River Tay valley, north of Dunkeld. The arable farming in the centre of the photograph is established on the alluvial soils and the humus-iron podzols of map unit 98 (Corby Association, Classes 3 and 4). The foreground shows the peaty podzols and humus-iron podzols of the moderately rocky map unit 506 (Strichen Association, Class 6). The brown forest soils, humus-iron podzols and humic gleys of map unit 505 (Strichen Association, Classes 4, 5 and 6) dominate the afforested area on the nearside of the river. The mainly afforested land on the far side of the river indicates the extent of map unit 498 (Strichen Association, Classes 3, 4 and 5), mainly humus-iron podzols, brown forest soils and gleys. In the far distance this map unit merges with map unit 505.* SDD Crown copyright.

188

Lomond Hills. There, freely drained brown forest soils formed in stony, sandy loam drifts of the Sourhope, Mountboy and Darleith Associations are well suited except for minor areas of occasional rock outcrops and steep rock-controlled mounds. Because the natural acid bent fescue and common white bent grasslands provide moderate grazings and are difficult to replace by surface treatments, reclamation is seldom carried out.

In the North-East Lowlands, Class 5.1 land is only of note in the Skene Lowlands where it occupies approximately 65 square kilometres mainly in Deeside; it is associated with freely drained brown forest soils formed in the moundy gravels of the Dinnet Association and with moderately coarse-textured brown forest soils of the Tarves Association on valley footslopes. It also occurs sporadically across the subregion on local summits with humus-iron podzols, rarely peaty podzols, of the Countesswells Association. Apart from minor pattern limitations, especially in the pitted outwash plain of the Muir of Dinnet, where only surface treatments would suit the shallow topsoils, and disregarding the separate problem of the moderately extensive afforestation, most areas would permit some flexibility, with conservation possible on the footslopes. Stoniness, including common boulders, would be a serious limitation to reclamation by ploughing in the relatively shallow soils of the Countesswells Association though the boulders could be windrowed prior to surface treatments. With a rainfall of less than 1000 millimetres and only moderate and moderately severe climatic constraints, poaching risks are low and stock have prolonged access.

West of Inverness and around the Moray Firth Lowlands and the adjacent foothills of the Northern Highlands, Class 5.1 land occupies approximately 65 square kilometres mainly below 300 metres on the higher slopes and exposed rock-controlled summits such as the spine of the Black Isle and the summit complex north of Dingwall. Other areas include the coastal ridges north of Rosemarkie and scattered patches between the Beauly Firth and Glen Urquhart. Outwith one small area on the south-facing slopes of Drummossie Muir, Class 5.1 land has not been mapped on the lowlands and surrounding foothills east of Inverness. All the above areas are dominated by freely drained humus-iron podzols with rare brown forest soils, generally shallow owing to rock, and occupying complex slopes. The main associations include those of Mount Eagle, Kessock, Braemore, Ethie, Arkaig and Orton. With a slightly higher rainfall, up to 1100 millimetres, and slightly more adverse conditions than in the North-East Lowlands and Central Lowlands, especially exposure, the agricultural emphasis is on grazing with conservation greatly restricted. Surface reclamation is the norm because of the stoniness, shallowness and slope pattern; approximately one third of the area is afforested.

Because of altitude coupled with severe to extremely severe climatic constraints, Class 5.1 land does not exist in the Cairngorm Mountains and the Monadhliath Mountains except for tiny areas of mainly free-draining brown forest soils, totalling approximately 20 square kilometres. These occur, for example, on soils of the Tarves Association at Corgarff in the valley of upper Donside and on subdued moraines of the Arkaig Association in Glen Garry and of the Strichen Association in Glen Girnaig east of Blair Atholl; they are frequently associated with abandoned crofts. A similar total area is distributed across the South-West Grampian Highlands where it is correlated mainly with brown forest soils and humus-iron podzols on moderately coarse-textured moraines around the shores and lower slopes of the major lochs; sometimes minor areas of alluvium and fluvioglacial sands and gravels are incorporated. Often the Class 5.1 land is the best available for holding ground and the siting of fanks.

Conservation is sometimes risked though the rainfall in the south-west often exceeds 1600 millimetres and usually restricts utilization to grazing.

The remaining area of approximately 120 square kilometres of Class 5.1 land in Eastern Scotland is concentrated in the foothills between Deeside and the Braes of Glenbervie west of Stonehaven, with outliers in upper Glen Esk, Strathardle and around the Lornty Burn north-west of Blairgowrie. Except for the freely drained brown forest soils of the Deecastle Association on slopes in Strathardle and those of the Strichen Association in some of the hummocky moraines around Lornty Burn, most soils are podzolic. They comprise freely drained humus-iron podzols and some peaty podzols of the Countesswells and Strichen Associations mainly on gentle and strong, non-rocky slopes. An area at Tarffside, Glen Esk, also includes sands and gravels of the Corby and Boyndie Associations in a complex of terraces and mounds surrounded by podzols of the Strichen Association on the nearby slopes.

Division 2   Land in this division occupies 2155 square kilometres (8.7 per cent of the land area) and is moderately suited to reclamation and use as improved grassland. Most of it experiences moderate to severe climatic constraints, though limited areas occur on the coast and are subject only to minor or no climatic restraints. Excepting the upper Spey and Forth valleys where the rainfall reaches 1600 millimetres, the bulk of Class 5.2 land receives less than 1200 millimetres, often only around 1000 millimetres. The soils are dominantly podzolic, with brown forest soils restricted to the Central Lowlands and upper Donside and Deeside. In the North-West and South-West Grampian Highlands, the moderately severe physical limitations involve both slope and pattern with the latter frequently associated with hummocky moraine and fluvioglacial outwash deposits. Elsewhere, excluding the coastal deposits, the division occurs mainly on long slopes, occasionally complex, which often pass into the subalpine soils zone. Rock is rarely a problem and wetness seldom becomes a serious limitation except (1) on gentle slopes, especially below spring lines, on till platforms and where induration and iron pans impede surface-water draining, and (2) in the higher rainfall areas. Climatic limitations, including exposure and shorter growing season, are allied to thickening organic topsoils and form the major restrictions. Under such conditions, the establishment of grassland is normally readily achieved though steep slopes, hummocky patterns and soil variability can lead to problems of maintenance. On the higher slopes and on slopes in the wetter areas where the soils commonly possess a thicker peaty topsoil, poaching can lead to deterioration of the sward and to invasion by mosses, particularly *Polytrichum spp.*, and by the heath rush, *Juncus squarrosus*. Nevertheless, successful reclamation is occasionally being established at ever-increasing altitudes in the relatively low rainfall areas. Recently, near Advie in Speyside, for example, within a rainfall of 900 to 1000 millimetres, approximately 40 hectares of a peaty podzol with iron pan on a slightly convex slope between 300 and 400 metres has been reclaimed to grassland from a grouse-moor. The land is only used by cattle, and some 90 suckler cows and their calves have been successfully summered for the past 3 years. Conversely, another project on a gentle slope below 300 metres rapidly regressed because of severe poaching following the failure to identify and disrupt an iron pan below gleyed upper horizons and above freely drained subsoil.

  Because of the long, non-rocky slopes and mainly moderate to severe climatic constraints, competition between forestry and agriculture is severe on Class 5.2 land. More importantly, however, most of the prime grouse-moors are also situated in the division with the vegetation of the podzolic soils being rotationally burned to ensure the dominance of heather.

Most of Class 5.2 land is correlated with the foothills fringing the North-East and North-West Grampian Highlands and the Northern Highlands where it occupies approximately 1355 square kilometres. The remainder is distributed mainly in the Spey valley, the deeply incised valleys of the South-West Grampian Highlands and in the lowlands. It is associated basically with podzols of the Strichen, Arkaig, Countesswells and the Foudland Associations; minor areas include the Tynet, North Mormond/Orton and Durnhill Associations. Brown forest soils are locally extensive in upper Donside and Deeside within the Insch and Tarves Associations and within the Deecastle Association in Glen Shee and around Blair Atholl. The slopes are mainly convex and free-draining, and the soils are often shallow owing to induration. Most soils are stony and there is frequently a pavement of stones at the organo-mineral interface which involves additional stone clearance where ploughing is used during reclamation.

Occasionally, at the upper limits where the peaty surface horizon thickens and the slopes rise towards the subalpine soils zone, the division 2 land merges with the division 3 land, for example, on the valley slope in the South-West Grampian Highlands and in the pitted outwash deposits in Speyside. More usually, however, the upper limits are marked by a distinct break in slope and a rapid change into Class 6.3 land on a planation surface.

Approximately 22 per cent of the Class 5.2 land occurs in the lowland subregions, much of it related to the upper slopes of isolated hills. In the North-East Lowlands, outwith the Skene Lowlands, it is restricted to podzols on such hills as Knockandy (Foudland Association) and Mormond Hill (Durnhill Association); in the Skene Lowlands, however, it is common on the slopes of most of the higher granite hills, including Bennachie, the Hill of Fare and Mortlich, as well as the lower slopes along Deeside south of Glen Tanar. Within the Central Lowlands, the division is located mainly on the central spine of the Sidlaw Hills (Sourhope and Mountboy Associations), the Ochil Hills (Sourhope Associations especially between Glen Devon and Glen Farg, and in the Lomond Hills (Darleith Association). Contrary to Class 5.2 soil conditions elsewhere in Eastern Scotland, most of these soils are brown forest soils supporting acid bent–fescue and common white bent grasslands; the natural, moderate grazing values often reduce the incentive for reclamation. The division also occurs on outlying and occasionally slightly rocky hills along the Highland Boundary Fault, involving mainly brown forest soils and humus-iron podzols of the Stonehaven, Callander/Gourdie/Strathfinella and Sourhope Associations. Along the Moray Firth Lowlands, part of the division is also closely associated with podzolic hill slopes, for example, around Monaughty Forest (Orton Association) near Elgin and the Ussie plateau (Halton Association) near Dingwall. In all the above situations, slope and pattern, coupled with increasingly adverse effects of exposure, are the primary limitations. It is also associated with the strongly ridged and often shallow ground on the gentle slopes of Drummossie Muir (Sabhail/Mount Eagle Associations).

Areas of peaty and noncalcareous gleys of the Balrownie Association in the upper Forth valley at Moor Park and around Loch Mahaick have responded well to intensive drainage schemes and are included in Class 5.2. Other minor areas of the division occur on the coast, mainly accommodating golf links, and are subject to limitations of erosion and drought.

Division 3 Land in this division is marginally suited to reclamation and use as improved grassland. It covers 1166 square kilometres (4.8 per cent of the land area) and is correlated primarily with soils having peaty surface horizons usually subject to wetness problems. A few minor exceptions are associated with low raised beaches, many of which are excessively drained; other minor areas have severe pattern limitations or steep slopes. The wet peaty areas, constituting the bulk of Class 5.3 land, are difficult to handle and usually necessitate the use of low ground pressure machinery and the installation of drains; often the timing of operations is severely restricted, though in at least one case in Speyside frozen conditions were made use of to successfully prepare the surface of a basin peat for seeding. Once established, the maintenance of sward and drains is normally difficult and expensive with the drains often short-lived due to ochre blockage and with the deterioration of the surface accompanied rapidly by reversion to the natural vegetation or by a change to rush-infested pasture, Poaching, although a constant and serious limitation, is mitigated in comparison with that in western Scotland by the much lower rainfall over most of the area, mainly below 1200 millimetres. Outwith the upper Forth valley where the rainfall around Flanders Moss reaches 1600 millimetres, a more flexible grazing system with both sheep and cattle is often possible, providing the cattle are judiciously rotated. The combined grazing habits usually result in a better and longer-lasting sward.

Within the lowland subregions where it occupies approximately 300 square kilometres, the division is distributed amongst four distinct landforms, namely peatlands, coastal deposits, freely drained upper slopes of isolated hills and wet lower slopes formed in till; most are non-rocky. The peatlands, mainly basins, range from moderate areas in the Buchan Platform west of St Fergus to the extensive Flanders Moss in the Forth Lowlands and the blanket peats on the eastern Braes of Doune; minor areas of peat and peaty alluvial soils occur in east Fife. Wet mineral soils are of limited extent and include the stony, sandy clay loams of the Hatton Association around Windyheads on the Upper Buchan Platform where much of the low ground has been reclaimed from abandoned, rush-infested grassland. Another area occurs in west Fife with fine-textured peaty gleys of the Giffnock Association forming a complex with small peat basins around Loch Glow in the Cleish Hills. Small areas of moderately fine-textured gleys, especially of the Balrownie Association, are found on lower slopes and depressions in the Forth Lowlands and Strathallan.

Class 5.3 land is a common feature of the non-rocky coasts especially on the southern shores of the Moray Firth and at the entrance to the Tay estuary. Textures are usually sandy and gravelly with landforms ranging from wet dune slacks to stabilized dunes. Drainage varies widely from excessively freely drained to very poorly drained. Wind erosion is a persistent hazard in the dune areas where the topsoil is usually so thin that only minimal surface treatments would be safe. Because of the historical problems of encroachment by aeolian sand on surrounding arable areas, many of these coastal deposits have been afforested, for example, Culbin, Lossie and Tentsmuir Forests. Some areas, too small for inclusion at the 1:250 000 scale, including renowned golf courses, have been amalgamated with the adjacent map unit. Most of the free-draining sands are quartzose and rapidly show signs of podzolization after afforestation.

On the isolated crests and upper slopes within the lowland subregions, the Class 5.3 land comprises mainly peaty podzols formed in a shallow, stony, sandy loam drift. Frequently an iron pan induces gleying in the overlying horizons which, in conjunction with moderately severe exposure, slope and pattern limitations, often makes reclamation an unattractive proposition. Examples include the Correen Hills (Foudland Association) in the Upper Buchan Platform, the Lomond Hills (Darleith Association) and the Ochil Hills (Sourhope Association) around the Water of May.

The bulk of Class 5.3 land occurs in the foothills and valleys of the Grampian Highlands and the Northern Highlands, covering approximately 860 square kilometres. About 25 per cent of that area occurs in the South-West Grampian Highlands where it is related principally to hummocky moraine of the Strichen Association in the long, narrow, valley floors and the adjacent valley slopes. There, the distinctive pattern of podzols on the mounds and gleys or peat in the hollows is the main limitation. Along the Highland Boundary Fault and in the Angus glens, however, the division is more closely associated with long, convex slopes supporting mainly peaty and humus-iron podzols and with concave, wet slopes supporting mainly peaty gleys. North of Dunkeld to Glen Esk, the area is dominated by soils of the Strichen Association where increasing altitude is accompanied by thickening peaty horizons and a rapidly shortening growing season; steep slopes are often additional limitations. South of Dunkeld, the topography is more complex and occasionally slightly rocky; the soils are still mainly peaty podzols and gleys and belong to the Strichen, Foudland, Callander/Gourdie and Stonehaven Associations. North-westwards from the headwaters in the Angus glens, little Class 5.3 land occurs, mainly because of altitude, until the middle and upper reaches of the River Spey and the River Findhorn. In these areas the division, though including long, steep, wet and dry slopes, is dominated by moundy outwash plains of the Corby and Boyndie Associations and the Dulsie Association. Pattern is the overriding limitation with the soils on the mounds almost exclusively podzols with iron pans, varying from humus-iron podzols to peaty gleyed podzols; basin peats, often extensive, usually fill the hollows. Many of the drainage problems in the podzols respond to deep tining because of the freely drained horizons below the iron pan.

Class 5.3 land occurs in the Grampian Foothills bordering the North-East Lowlands probably because of the combination of the dominantly convex slopes, the mainly highly porous and shallow stony sandy loam drifts and the relatively low rainfall of less than 1000 millimetres. Westwards from Darnaway Forest, however, the division reappears in the foothills and peneplain fringing the Moray Firth Lowlands where gentle slopes have accelerated the development of thick peaty surface horizons, for example around Loch Ashie on Drummossie Muir. Other areas occur on the steeper, complex slopes of the foothills of the Northern Highlands west of Inverness. The soils are mainly peaty podzols, except on the gentle slopes where peaty gleys dominate, and belong mainly to the North Mormond/Orton, Arkaig and Kindeace Associations.

Class 6

Class 6 land is unsuited to mechanical improvement but its associated vegetation is capable of producing sustained grazing values. It covers 8017 square kilometres in Eastern Scotland (32.5 per cent of the land area) and is overwhelmingly, but not completely, correlated with altitude; it dominates the North-East, North-West and South-West Grampian Highlands and the Northern Highlands where it is subjected to very severe climatic constraints. Only 385 square kilometres are recorded in the lowlands, mainly in the Ochil Hills and along the coast.

The productivity of Class 6 land is reflected primarily by the natural vegetation which varies from the relatively highly productive communities such as the bent–fescue and meadow-grass–bent grasslands to the poor, mainly ericaceous and acidophyllous communities such as heather moors and blanket bogs. The vegetation, of course, reflects the influence of soil, landform and altitude. Today, the dominant domestic grazing animal is the sheep with the traditional summering of cattle on the high-level pastures in Eastern Scotland long since abandoned; cattle in general on Class 6 land no longer figure prominently, their numbers being dictated by the amount of arable land and 'green' grassy hills within the farm. With regard to sheep, not only are stocking rates and breed determined by the grazing quality but also the size and type of sheep carried. Thus, Blackface are more common on the heather-dominated hills whereas Cheviots are typical on the green western and southern hills. Stocking rates vary enormously though they are usually low in the high deer-forests in the Monadhliath Mountains. Because of the relatively low productivity of even the better, natural communities compared with improved pasture, the farming is one of low input. For example, new fencing is often restricted to deer fences protecting low ground and forests, the original Victorian iron post and wire fences now totally neglected.

The class is divided into three divisions based upon the plant communities present and the evaluation of their palatability, digestibility and productivity. An explanation of the method of interpretation of the grazing values is given in Bibby *et al.*, (1982).

Division 1   This land contains vegetation with high grazing values which in Eastern Scotland are associated mainly with bent–fescue grasslands and, more rarely, rush pastures. It occupies only 270 square kilometres of which 50 per cent is found along the floors and steep sides of the glacially overdeepened valleys in the South-West Grampian Highlands. There, it is common in Glens Dochart and Lochay, the Braes of Balquidder and in the districts of Strathyre and the Trossachs. Situated in an area of high rainfall of 1400–2400 millimetres, most of the soils are brown forest soils and, to a lesser extent, humus-iron podzols of the Strichen Association; minor areas are occupied by the Foudland and Stonehaven Associations. These soils are mainly freely drained, stony sandy loams developed in colluvium and on hummocky moraine on the valley floor and adjacent lower slopes. The moderately coarse texture, allied to the steep slopes, promotes a rapid run-off, facilitating free drainage and inhibiting the development of peaty surfaces. As in many similar situations to the west and south, bracken is an ever-increasing weed problem whose rhizomatous rooting system effectively exploits the freely drained loose fabric of the subsoil. This weed canopy severely reduces the productivity of the shaded grasses as well as being poisonous to cattle under certain circumstances.

Outwith the South-West Grampian Highlands, the only significant area of Class 6.1 land occurs on the scarps of the Ochil Hills where the soils are mainly brown forest soils with some brown rankers and gleys of the Sourhope Association. Smaller areas occur on Mile Hill north-west of Kirriemuir and on Pressendye Hill north of Tarland where the division has been recorded on peaty podzols of the Tarves Association; it has been recorded also on podzols of the Strichen Association in Glen Esk. Minor areas occur on limestone soils, for example, in Glen Tilt and Glen Shee but are too small to identify at the 1:250 000 scale and have been included in adjacent Class 6.2 land as a mosaic of high and low grazing values.

Division 2   Class 6.2 land covers 1870 square kilometres (7.6 per cent of the land area) and is related to plant communities with moderate grazing values such as common white bent and flying bent grasslands, common cotton-grass bogs and stiff sedge–fescue grassland or to a mosaic of communities with varied values but aggregating a moderate value. It is restricted largely to the Grampian Highlands and Northern Highlands where it occupies steep valley sides, long, non-rocky slopes mainly in Angus, and high-level pasture of the subalpine and alpine soil zones dominated by snow-bed communities.

Although the vegetation of the mountains in Eastern Scotland is primarily acidophyllous and ericaceous, and acidic parent materials and podzolization are dominant, there are regional trends as well as local variations which encourage the development of plant communities with moderate grazing values. In particular, the ericaceous dominance is increasingly replaced westwards and south-westwards by *Nardus stricta*, *Eriophorum vaginatum* and *Molinia caerulea* with these accompanying a rapidly rising rainfall and a commensurate flushing in the soils. In addition, in the more strongly dissected South-West Grampian Highlands, colluviation is an important soil-forming process which in the milder climate promotes the development of brown forest soils and grasslands, especially in the widespread and relatively more base-rich parent materials derived from rocks of the Dalradian Assemblage. The common change westwards and southwards from coniferous to broadleaved woodlands, mainly birch and oak, is also notable and facilitates the replacement of heath species under the canopy on the lower and mid slopes.

In the South-West Grampian Highlands, Class 6.2 land covers some 580 square kilometres and is confined mainly to the steep and very steep slopes of the over-deepened valleys. There, it comprises the brown forest soils on the lower slopes, sometimes including the hummocky moraine but often lying immediately upwards of the Class 6.1 land and Class 5 land so characteristic of the valley floors and lowest slopes. Often under broadleaved woodlands and bent–fescue grassland, the brown forest soils are replaced upslope by peaty podzols supporting heath rush–fescue grasslands. In places, these podzols merge into subalpine soils with upland bent–fescue grasslands and stiff sedge–fescue grasslands but, above about 600 metres, these subalpine soils under the wetter, more exposed conditions are mainly peaty gleys and peat with bog heather moors and mountain blanket bog communities which form Class 6.3 land. Around Achray and Loch Ard Forests, the division atypically occupies non-rocky, lower ground with gentle to steep slopes where humus-iron podzols and some brown forest soils of the Strichen and Foudland Associations support acid bent–fescue grasslands; at Achray and elsewhere, peaty gleys of these associations support rush pastures. Class 6.2 land also includes soils of the Arkaig Association on hummocky moraines, for example in Glen Garry and around Loch Rannoch.

In the North-East Grampian Highlands, Class 6.2 land covers some 635 square kilometres and much of it is concentrated across the south-eastern slopes especially between Glen Clova and Glen Esk where humus-iron podzols of the Strichen Association supporting acid bent–fescue grasslands and boreal heather moors gradually merge upwards into peaty podzols carrying upland bent–fescue grasslands and lowland and upland blanket bogs. These slopes are essentially non-rocky and form the well-known green braes of Angus, long famous for sheep farming. Class 6.2 land is also associated with the headwaters of the rivers flowing south-eastwards through such glens as Prosen, Clova, Glenshee and Isla. In these rocky, higher sites, usually above 600 metres, the soils are mainly peaty podzols of the Tarves Association supporting white bent grasslands and succeeded upslope by subalpine and alpine soils carrying upland bent–fescue and stiff sedge–fescue grasslands.

In the Monadhliath Mountains, the Class 6.2 land occupies about 350 square kilometres both in deeply entrenched river valleys, such as those of the Findhorn and Killin, and on high-level, mostly freely drained, subalpine and alpine soils of the Arkaig and Countesswells Associations. The alpine soils are associated with characteristic snow-bed communities including stiff sedge–fescue grasslands, viviparous fescue grasslands and mountain heath rush grasslands. These form the traditional 'summerings' without which sheep farming in the Monadhliath Mountains would be difficult because the surrounding blanket bog communities on the hagged peat are little used except for the sheep 'runs' along the flushed hags. The high-level pastures are intensively used by the red-deer population with both deer and sheep using the deeply incised valleys for shelter in bad weather. Towards the south-west, between the Corrieyairack and Loch Ericht, peaty gleys become widespread, supporting acid bent–fescue grassland, white bent grassland and flying bent grassland.

Without the availability of low ground for shelter and wintering, let alone the provision of winter feed, all these high-level pastures and much of the less productive mid-ground would be untenable. Indeed, but for the fortuitous access and low ground provided by Strathspey and Strath Tay, much of the central Grampian Highlands would not be economically viable for sheep under present-day conditions. Outwith the mountain regions, Class 6.2 land occupies only approximately 200 square kilometres, of which the greater area is associated mainly with the white bent grassland on peaty and humus-iron podzols of the Sourhope Association on the moderately rocky, upper slopes of the Ochil Hills. A similar situation obtains on the stepped, northern flanks of the Gargunnock Hills where brown forest soils of the lower slopes carry bent–fescue grassland but are replaced by peaty podzols with white bent grassland on the steeper, upper slopes. Minor areas also occur on peaty podzols, mainly of the Countesswells Association, on the largely non-rocky, upper slopes of isolated hills in the North-East Lowlands such as the Hill of Fare and Bennachie. Most of the remaining areas of Class 6.2 land in the lowland subregions occur along the coast from north of Aberdeen to the Black Isle. These are very diverse in nature and range from the regosols of the Links Association at Culbin and the Sands of Forvie to the undifferentiated raised beaches of the Nigg Association such as the shingle beaches near Lossiemouth; also included are humus-iron podzols and, to a lesser extent, brown forest soils on the steep rock-controlled slopes and talus of the Kessock and Ethie Associations on the Black Isle coastline.

Division 3   Covering 5877 square kilometres (73 per cent of Class 6), this division is the most extensive land type in Eastern Scotland, occupying about 25 per cent of the region. It is widespread across the Grampian Highlands and the Northern Highlands where it is dominant between 500 and 1000 metres within a rainfall which varies from 1200 to 3200 millimetres. The vegetation, though comprising a wide range of species, is dominated by acidophyllous plants with low grazing values. Plant communities are mainly heather moors, deer-grass and cotton-grass bogs, blanket bogs and mountain heaths. The soils are acid and podzolic with peaty surface horizons though gleys are common towards the west, and subalpine soils at the highest altitudes; organic soils are frequent in the North-West and North-East Grampian Highlands where the planation surfaces favour their development. In addition to the Organic Soils, the Class 6.3 land is dominated by the Arkaig, Countesswells, Strichen and Tarves Associations.

Only about 200 square kilometres of this division have been recorded outwith the Grampian Highlands. Most of this area is located in the Central Lowlands on the peat-covered summits of the Ochil Hills and Gargunnock Hills. Smaller areas include the slightly rocky summits of the Lomond Hills (Darleith Association) and the rocky summits of isolated hills in the North-East Lowlands such as Bennachie (Countesswells Association) and Knock Hill (Durnhill Association); also included are the shingle deposits and dune sands of Culbin Forest in the Moray Firth Lowlands.

Sheep stocking densities are low and occasionally the hirsels have been totally displaced in favour of red deer in the more isolated and higher deer-forests in the North-East and North-West Grampian Highlands. Primary sporting interests, high lambing mortality and disease, especially tick-borne, are the principal factors in this displacement. Management is usually restricted to rotational burning of the widespread heather moors and bogs and to more frequent burning of the flying bent and white bent pastures in the wetter western and south-western areas.

Class 7

Class 7 land covers 1636 square kilometres (6.7 per cent of the land area). The limitations are extremely severe and cannot be rectified.

Much of this land in Eastern Scotland consists of exposed summits and plateaux in the alpine soils zone where climate is the dominant limiting factor

Plate 20. *Caenlochan, south of Braemar, well known for its rare plants. The moraines with peaty podzols of map unit 504 (Strichen Association) are prominent in the valley floor; on the slopes in the foreground there are peaty podzols of map unit 499 (Strichen Association). Grazings on both units are mainly of low value (Class 6.3). On the rock wall to the left are rankers and lithosols of map unit 515 (Strichen Association), and at the head of the valley and to the right are the alpine soils and peat of map units 256 and 257 (Foudland Association). On the plateau of Glas Maol above there are extensive areas of alpine soils, map unit 514 (Strichen Association). These map units are mainly on Class 7, though there are some grazings of moderate and low value on the steep slopes. Such slopes provide good shelter for red deer.* Cambridge University Collection: copyright reserved.

197

(Plate 20). Lithosols and rankers are widespread and extremely rocky areas, including major screes, are common. Vegetation is normally sparse, wind-cropped owing to extreme exposure and of negligible grazing value except in locally sheltered snow-beds where *Nardus stricta, Deschampsia flexuosa* and *Carex* spp. are often present and provide grazing for a limited season. The main areas occur on the Cairngorm Mountains, the Ben Alder and Creag Meagaidh ranges and Ben Wyvis. Minor areas are typical of most isolated peaks above 800 metres in the South-West Grampian Highlands.

Across much of the Monadhliath Mountains and parts of the North-East Grampian Highlands, peat mantles the plateaux between 700 and 900 metres; around Tomatin the altitude limit falls to 500 metres. Whereas many other peatlands are mapped within Class 6.3, severe hagging is so dominant in these areas that they are restricted to Class 7. In addition to the extremely severe climatic restraints, the broken ground creates severe problems in stock control especially when late-lying snow fills the bogs until early summer. Erosion in limited areas has resulted in unvegetated, redistributed peat.

Outwith the mountains, Class 7 land has been recorded only in the coastal deposits, including the constantly evolving off-shore bars at Nairn and the dune complex at Barry Links east of Dundee.

LAND CAPABILITY FOR NON-AGRICULTURAL USES

Whereas agriculture dominates the lowlands and much of the hill ground, many other uses exert varying demands on the land; these include forestry, building and road construction, various extractive industries and recreation. In Eastern Scotland, forestry, agriculture and grouse-shooting are often competing forms of land use, especially in the drier eastern areas; in the mountains and plateaux, deer-stalking and sheep-grazing are dominant although recreational activities such as skiing, hill-walking, climbing, camping and caravanning, are causing increasing pressures on the land resource. Many of the factors considered in the evaluation of land for recreation and other non-agricultural uses are reviewed by Bartelli *et al.* (1966), Jarvis and Mackney (1979) and McRae and Burnham (1981); those concerning forestry have been examined amongst others by Page (1970) and McCormack (1970).

FORESTRY

Many centuries ago, Virgil observed that 'not every land can nourish every tree' (Stone 1961), a comment highly relevant to Eastern Scotland where trees vary from the giants of Murthly Castle policies to the stunted specimens of *Betula nana* and *Salix herbacea* on the highest plateaux. In general, forestry in the lowlands occupies only land where soil and physiography restrict the value of agriculture and where some estates have developed and maintained forests for centuries. The natural climax species on the podzolic soils north of the Highland Boundary Fault is the Scots pine with large remnants of the Caledonian pinewoods surviving at Abernethy and Rothiemurchus in Speyside. Excluding birch, natural deciduous woodlands are usually associated with brown forest soils and are fairly rare north of the Fault, though excellent examples of oak and mixed woodland occur in sheltered and strongly colluviated soils at Killiecrankie,

198

Glen Tarff and between Beauly and Garve. Birch, however, is widespread but is probably best developed in the wetter areas around Lochs Laggan and Ness.

Tree growth is affected by a number of factors which include, either singly or in combination, climate, shallowness owing to rock and induration and wetness. In Eastern Scotland, unfavourable climate, resulting from exposure allied to temperature and related to altitude, is the major restricting factor. This is well illustrated by the highest surviving natural tree line in Scotland at the entrance to Glen Einich. There, Scots pines, accompanied by a juniper scrub belt, decrease rapidly in height upon entering the subalpine soils zone at around 600 metres. Birch displays a similar pattern on Morrone near Braemar at approximately the same height. Commercial forestry, however, rarely exceeds 500 metres and often shows stunting at the highest levels. Depending upon surrounding topography, isolated summits at much lower levels can also display severe stunting though the canopy just below the top can be approaching normal height, for example, Carn a Chnuic.

Exposure coupled with shallow rooting systems often induces windthrow, the limitation restricting management through a shortened crop rotation and consequently a reduced pole height. Shallowness results from a variety of factors including a high, poorly oxygenated ground-water table, an indurated horizon especially where accompanied by surface-water gleying, finer-textured subsoils and rock; occasionally surface water, induced by an iron pan, is responsible though where freely drained subsoils exist below the pan most tap roots are capable of penetrating the barrier. The influence of drainage upon shallowness is often demonstrated by windthrow being restricted to peaty gleys with indurated subsoils despite surrounding freely drained podzols having similar induration. Solid rock is obviously restrictive but rooting can be deep in freely drained, shattered rock and windthrow is seldom a problem. Very stony horizons in shallow soils can also alter potential stability by reducing the available rooting volume.

Available soil nutrients are broadly related to the soil rooting volume qualified by the soil parent material, the major soil subgroup and the intensity of gleying, leaching and flushing. Species site and nutrient requirements, however, vary widely and crop performance may be more closely linked to other site factors than to inherent mineralogy though there is a general trend for growth to decrease from basic to acidic soil parent materials. Frequent attempts have been made to predict site capability using selected site factors within regressive equations but with varying conclusions (Page, 1970; Cook et al., 1977). Thus, the latter claim 'that the high inherent variability of the Scottish uplands precludes the development of prediction equations of sufficient accuracy which can be applied over a reasonably extensive area but that a physiographic approach to land valuation may be more relevant to the needs of forestry.'

Apart from climate, the principal factors limiting forestry in much of Eastern Scotland are induration restricting rooting volume, and heather suppressing many species, especially Sitka spruce. Both problems are ameliorated by cultivation with deep tine-ploughing to disrupt impermeable layers a standard practice. Using tracked vehicles and mounted ploughs, one-way ploughing is possible on non-rocky slopes up to about 30 degrees; wetness of slope, degree of rockiness and presence of boulders all progressively increase difficulty of operation. After planting, it is essential to apply phosphate on non-flushed peatlands with the poorest sites also requiring potash and possibly nitrogen; all peatlands benefit from topping up with phosphate and potash after a few years. Phosphate is also essential for Sitka spruce when planted on a moorland dominated by Calluna, and is often used to bring trees out of check as well as being a routine application at planting on most acid soils. Severe phosphate deficiency on soils derived from basic igneous rocks (Darleith Association) and ultrabasic rocks (Leslie Association) was found to be correlated apparently with high amounts of amorphous aluminium which may also inhibit mycorrhizal development (James et al., 1978). Nitrogen deficiency in coniferous crops, including Corsican pine at Culbin and Scots pine on Deeside, is apparently correlated with excessive immobilization of nitrogen in the mor humus of podzols. These have a low nitrogen capital and vary from incipient podzols in aeolian sands to humus-iron podzols in fluvioglacial gravels; all are excessively drained and liable to drought.

Because of the variety of sites and the prevalence of dry conditions, a wide range of species has been planted. Nowadays, Sitka spruce, owing to its high growth yield, is grown wherever moisture reserves are adequate with Scots pine favoured on the drier heather moors. The more demanding species such as European larch, western red cedar and Douglas and grand firs are still planted on sites of higher fertility whereas lodgepole pine is well suited to peatlands and low fertility sites as is Corsican pine to the coastal sands and gravels.

RECREATION AND OTHER LAND USES

Skiing is a major expanding activity in the region, geared largely to slalom and downhill skiing with centres at Coire Cas, Cairnwell and the Lecht but with cross-country skiing developing rapidly. Growing pressure on extant resources has already led to surveys to identify other suitable areas and, coupled with public concern over the erosion and other problems associated with the creation of these centres, has led to the production of guidelines by the Scottish Development Department for the future development of skiing areas. Planning application for expansion at the Lecht is under review and soil survey data have been used to identify potential erosion and assist planning decisions; data were also provided for the public enquiry into the proposed expansion at Lurchers' Gully, Cairngorm. Many of the original erosion problems stemmed from the failure to recognize the fragile balance of the soils and plants relationship in the subalpine and alpine soils zone. Much effort and some research have followed in attempts to colonize bare erosion patches though these have been based mainly on commercial grass mixtures and the use of various bitumastic sprays and mulches; they have been only partially successful. Skiing-centre management is keenly aware of the erosion problems caused by construction at high altitudes and which automatically follow ground disturbance at these levels; in recent developments, for example at Coire na Ciste, erosion was greatly minimized by off-site fabrication and uplift by helicopter. Apart from construction scars, small-scale erosion is initiated by damage to vegetation, both on and off the pistes, by cutting from ski-edges following partial snowmelt. Because of the even more fragile environmental balance on the highest plateaux, especially related to the weakly developed structure of alpine soils and the associated patchy vegetation, any major expansion in cross-country skiing along specific routes would require careful monitoring.

The substantial expansion of hill roads in recent years has been visibly accompanied by erosion in many instances. Although some are grant-aided for agricultural purposes and all require planning permission above 300 metres, most have been developed for the shooting — grouse and red deer and, in some areas, ptarmigan and mountain hares. Occasionally, the gradient is the only consideration afforded the route and sometimes little understanding is shown of the relationship between soils and drainage. Gully erosion and minor landslips are all too common, the inevitable outcome of low-cost installation with inadequate drains and culverts. Peat, peaty gleys with indurated subsoils and peaty podzols with gleying above an iron pan are particularly liable to damage.

Soil maps are of paramount importance in the delineation of areas of sand and gravel deposits (Chester, 1980) and the 1:250 000 series provides the first national map coverage of the resource. Classified by sieve sizes, sand and gravel form the most widely used source of aggregate and constitute about 47 per cent of the total aggregate consumption of 12 million tonnes in Great Britain (Scottish Development Department, 1977).These deposits are particularly important in Eastern Scotland where the bulk is found in the Corby and Boyndie Associations; for example, within Grampian Region approximately 2.5 million tonnes have been extracted annually since 1975. Although soil survey data do not normally allow the quantification of the deposits, they do determine the areal distribution and summarize the major lithological constitutents, the latter vital factors for concrete manufacture.

Peatlands, whose total area has been variously reported as between 710 000 and 833 000 hectares, form a major national resource which is well served by the soil map. Although most lowland sites have been well documented by the Peat Survey and much information is contained in maps produced at 1:63 360 by the Soil Survey and the Geological Survey, the national distribution of peatlands is objectively presented for the first time. In Eastern Scotland, many of the basin peatlands are currently exploited for fuel and horticultural purposes though many more were completely exhausted in the past or removed during agricultural reclamation. Increasingly, conservation is being advocated as their primary use.

Field sport, particularly the shooting of grouse and red deer, forms a major land use with some estates in the central Grampian Highlands devoted entirely to sport. Despite the controversy surrounding grazing competition between sheep and deer, and deer having a better browsing ability, most feeds used for farm animals have been tried successfully in deer-farms (Scottish Agricultural Colleges, 1979). It is possible therefore that the relative grazing values established for sheep in Class 6 land are relevant to deer. The lack of competition or reduced competition, owing to displaced and reduced hirsels on the better quality grazing, especially the snow-bed communities and the grasslands on steep-sided valleys, has probably contributed to the present expansion of the national red deer herd to approximately 270 000. Currently, venison sales and stalkings contribute £1.25 million annually, though population pressure is increasing damage to young Sitka spruce and lodgepole pine plantations and to marginal farming areas, mainly Class 4 land in the valley floors.

In about 15 years, deer-farming has come almost full circle from its initial concept of utilizing the rough grazings of Class 6 land by means of open hill paddocks to a park system. Nowadays, most of the 50 deer farms are widely distributed throughout Scotland except the north-west and occupy relatively low ground, mainly Class 4 land. Though they contribute only a fraction compared with the 2500 tons of wild venison produced annually, deer farms are steadily evolving and assimilating the results of research pursued by various government organisations such as the Rowett Institute and the Hill Farming Research Organization at Glensaugh experimental farm.

Red grouse is directly linked with the staple diet of heather and therefore figures prominently on the dry podzolic slopes of eastern Scotland where it often plays the dominant role in the economy and interest of many estates. Although presently suffering a marked decline in numbers, the grouse provide an income from prime moors up to £25 000 rental in the first week of the shooting season. Grouse are both walked-up and driven. Closely associated with the sport is the rotational burning, a systematic practice developed only two centuries ago following the introduction of sheep, though natural and deliberate forest fire clearances probably initiated the development of the moorlands. Repeated burning has gradually reduced floristic variety and produced either heather-dominated moors or poor grassland often characterized by *Nardus stricta*, depending upon the grazing and stocking rates. There is evidence that the healthiest and most efficient breeding stocks are associated with podzols derived from base-rich parent materials. In the wetter, western areas, heather is often completely displaced by *Molinia caerulea Nardus stricta* and *Eriophorum vaginatum* by the large annual spring fires designed to create an early 'bite' in the more grassy vegetation.

 Other major uses of land include building and road construction. Apart from identifying sand and gravel deposits, the soil map can be used to locate suitable soils for infill and embankments. The distribution of soils is also an important consideration in the evaluation of alternative routes in initial planning stages; soil and peat survey data were used in the determination of subsoil conditions of the new A9 trunk road between Dalwhinnie and Daviot. Within the region, several costly remedies have followed the failure to identify potential problems evidenced by soil morphology during building-site soil investigations. These usually involved ground-water and alluvium.

Pressure from certain recreational pursuits can be predicted readily in many areas where the environment is keenly balanced. Thus, links, dunes and raised beaches which form primary areas for camping, caravanning and picnicking are particularly prone to major erosional problems because of the weak structures in the topsoils and single grain subsoils. The degeneration of these coastal deposits, resulting from greatly increased use, has led to a series of reports prepared for the Countryside Commission (Ritchie *et al.*, 1969, 1970) by the Geography Department, University of Aberdeen; the latter had previously documented the evolution and condition of many beaches and drawn attention to the inherent problems following misuse. Though many coastal areas in Eastern Scotland are not generally available to the public because large areas are afforested, or under the control of the Ministry of Defence, or are protected as conservation sites, a few problems have developed in dune areas, for example, at Balmedie and Findhorn. Hill-walking constitutes another pursuit in which excessive trafficking rapidly induces erosion. Where the alpine and subalpine soil zones are involved, degradation is accelerated because of the thin organic turf, highly susceptible to treading damage and because of the poorly developed and loose structure in the upper horizons. Paths on Cairngorm and Ben Lawers provide excellent examples of excessive wear and tear following their promotion as tourist attractions; the damage has been thoroughly researched by many organizations though solutions are less easily produced. Peaty soils at lower levels are also highly susceptible to treading damage.

References

Armstrong, M. (1977). The Old Red Sandstone of Easter Ross and the Black Isle. The Moray Firth Area Geological Studies. Inverness: Inverness Field Club.

Bartelli, L. J., Klingebiel, A. A., Baird, J. V. and Heddleson, M. R. (1966). Soil surveys and land use planning. Madison, Wisconsin: Soil Science Society of America and American Society of Agronomy.

Bibby, J. S., Douglas, H. A., Thomasson, A. J. and Robertson, J. S. (1982). Land Capability Classification for Agriculture. Monograph. Aberdeen: The Macaulay Institute for Soil Research.

Birse, E. L. (1971) Assessment of climatic conditions in Scotland. 3. The bio-climatic sub-regions. Aberdeen: The Macaulay Institute for Soil Research.

Birse, E. L. (1980). Plant communities of Scotland. Revised and additional tables. Aberdeen: The Macaulay Institute for Soil Research.

Birse, E. L. (1982). The main types of woodland in North Scotland. *Phytocoenologia*, **10**, 9–55.

Birse, E. L. and Dry, F. T. (1970). Assessment of climatic conditions in Scotland. 1. Based on accumulated temperature and potential water deficit. Aberdeen: The Macaulay Institute for Soil Research.

Birse, E. L. and Robertson, L. (1970). Assessment of climatic conditions in Scotland. 2. Based on exposure and accumulated frost. Aberdeen: The Macaulay Institute for Soil Research.

Birse, E. L. and Robertson, J. S. (1976). Plant communities and soils of the Lowland and Southern Upland Regions of Scotland. Aberdeen: The Macaulay Institute for Soil Research.

Bremner, A. (1942). The origin of the Scottish river system. *Scott. geogr. Mag.*, **58**, 15–20, 54–9, 99–103.

Chester, D. K. (1980). The evaluation of Scottish sand and gravel resources. *Scott. geogr. Mag.*, **96**, no. 1, 51–62.

Clapham, A. R., Tutin, T. G. and Warburg, E. F. (1962). *Flora of the British Isles* (2nd edn). London: Cambridge University Press.

Charlesworth, J. K. (1955). The late-glacial history of the highlands and islands of Scotland. *Trans. R. Soc. Edinb.*, **62**, pt. 3, no. 19, 769–928.

Cook, A., Court, M. N. and Macleod, D. A. (1977). The prediction of Scots pine growth in north-east Scotland using readily assessable site characteristics. *Scott. For.*, **31**, 251–64.

Fleet. H. (1938). Erosion surfaces in the Grampian Highlands of Scotland. Rapp. Comm. Cartog. des Surf. D'Appl. Text., Union geogr. Internat., 91–4.

Fraser, G. K. (1954). Classification and nomenclature of peat and peat deposits. *Proc. Int. Peat Symp.*, Dublin Sect. 13, 2.

Gauld, J. H. and Bell, J. S. (1983). Soils and associated management problems within the carses of Gowrie and Earn, Perthshire. *Scott. geogr. Mag.*, **99**, no. 2, 77–88.

Geikie, Sir. A. (1901). *The scenery of Scotland.* London: Macmillan.

George, T. N. (1965). The geological growth of Scotland. In: *The geology of Scotland* (ed. G. Y. Craig), pp 1–48) Edinburgh: Oliver and Boyd.

Glentworth, R. (1954). The soils of the country round Banff, Huntly and Turriff. (Sheets 86 and 96). *Mem. Soil Surv. Scot.* Edinburgh: HMSO.

Glentworth, R. and Muir, J. W. (1963). The soils of the country round Aberdeen, Inverurie and Fraserburgh. (Sheets 77, 76 and 87/97). *Mem. Soil Surv. Scot.* Edinburgh: HMSO.

Glentworth, R., Mitchell, W. A. and Mitchell, B. D. (1964). The red glacial drift deposits of north-east Scotland. *Clay miner. Bull.*, **5**, 373–81.

Green, F. H. W. (1974). Climate and weather. In: *The Cairngorms, their natural history and scenery* (eds D. Nethersole–Thompson and A. Watson), pp. 228–36. London: Collins.

Harris, A. L. (1977). Metamorphic rocks of the Moray Firth district. Moray Firth Area Geological Studies. Inverness: Inverness Field Club.

Harris, A. L. and Pitcher, W. S. (1975). The Dalradian Supergroup. In: *A correlation of the Precambrian rocks in British Isles*: (eds Harris, A. L., Shackelton, R. M., Watson, J., Downie, C., Harland, W. B. and Moorbath, S.) pp. 52–75. Spec. Rep. Geol. Soc. London, 6.

Hinxman, L. W. and Anderson, E. H. (1915). The geology of Mid-Strathspey and Strathdearn. *Mem. geol. Surv. Scot.* Edinburgh: HMSO.

Hulme, P. D. (1980). The classification of Scottish peatlands. *Scott. geogr. Mag.*, **96**, no. 1, 46–50.

James, H., Court, M. N. and Macleod, D. A. (1978). Relationships between growth of Sitka spruce (*Picea sitchensis*), soil factors and mycorrhizal activity on basaltic soils in western Scotland. *Forestry*, **51**, no. 2, 105–19.

James, P. W. (1965). A new check-list of British lichens. *Lichenologist*, **3**, 95.

Jarvis, M. G., and Mackney, D. (eds) (1979). Soil survey applications. Soil Surv. Tech. Mono., no. 13. Harpenden: Soil Survey of England and Wales.

Johnstone, G. S. (1966). The Grampian Highlands (3rd edn). *Br. reg. Geol.* Edinburgh: HMSO.

Kennedy, W. Q. (1946). The Great Glen Fault. *Q. Jl. geol. Soc. Lond.* **102**, 41–77.

Laing, D. (1976). The soils of the country round Perth, Arbroath and Dundee. (Sheets 48 and 49). *Mem. Soil Surv. Scot.* Edinburgh: HMSO.

Linton, D. L. (1951). Problems of Scottish scenery. *Scott. geogr. Mag.*, **67**, 68–85.

Manley, G. (1971). Mountain snows of Britain. *Weather*, **26**, 192–200.

McCormack, R. J. (1970). The Canada Land Inventory. Report No. 4. Land capability classification for forestry. Ottawa: Department of Regional Economic Expansion.

McRae, S. G. and Burnham, C. P., (1981). Land evaluation. (Monographs on soil survey.) Oxford: Clarendon Press.

Ogilvie, A. G. (1923). The physiography of the Moray Firth coast. *Trans. R. Soc. Edinb.*, **53**, 377–404.

Page, G. (1970). Quantitative site assessment: some practical applications in British forestry. *Forestry,* **43,** no. 1, 45–56.

Paterson, I. B. (1974). The supposed Perth Readvance in the Perth district. *Scot. Jl. geol.,* **10,** 53–66.

Peach, B. N. and Horne, J. (1930). *Chapters in the geology of Scotland.* London: Humphrey Milford.

Phemister, J. (1960). Scotland: The Northern Highlands (3rd edn.), *Br. reg. Geol.* Edinburgh: HMSO.

Piasecki, M. A. J. and van Breemen, O. (1983). Field and isotopic evidence for a *c.*750 Ma tectonothermal event in Moine rocks in the Central Highland region of the Scottish Caledonides. *Trans. R. Soc. Edinb., Earth Sciences,* **13,** 119–34.

Ragg, J. M., Bracewell, J. M., Logan, J. and Robertson, L. (1978). Some characteristics of the brown forest soils of Scotland. *J. Soil Sci.,* **29,** no. 2, 228–42.

Read, H. H. and MacGregor, A. G. (1948). The Grampian Highlands (2nd edn). *Br. reg. Geol.* London: HMSO.

Ritchie, W. (1972). The evolution of coastal sand dunes. *Scott. geogr. Mag.,* **88,** 19, 35.

Robertson, I. M. (1933). Peat Mosses. Pt. 1. Their development and early utilization in Scotland. *Scott. Jl. of agric.,* XVI 50–8, 160–72, 327–35, 467–72.

Ross, S. (1976). Climate. In: *The Moray Book.* (ed. D. Omand), pp 28–48. Edinburgh: Paul Harris Publishing.

Scottish Agricultural Colleges (1979). Red deer farming. The Scottish Agricultural Colleges, publn. no. 50.

Scottish Development Department (1977).

Simpson, J. B. (1933). The late-glacial readvance moraines of the highland border west of the River Tay. *Trans R. Soc. Edinb.,* **57,** 633–45.

Sissons, J. B. (1963). The Perth Readvance in central Scotland. *Scott. Geogr. Mag.,* **76,** 131–46.

Sissons, J. B. (1976). *The geomorphology of the British Isles: Scotland.* London: Methuen.

Small, A. and Smith, J. S. (1971). The Strathpeffer and Inverness Area. British landscapes through maps (13). K. C. Edwards. Sheffield: The Geographical Association.

Smith, T. E. (1968). Tectonics in upper Strathspey, Inverness-shire, Scotland. *Scott Jl. geol.,* **4,** 68–84.

Smith, D. I. (1977). The Great Glen Fault. The Moray Firth Area Geological Studies. Inverness: Inverness Field Club.

Smith, J. S. (1977). The last glacial epoch around the Moray Firth. The Moray Firth Area Geological Studies. Inverness: Inverness Field Club.

Smith, A. J. E. (1978). *The moss flora of Britain and Ireland.* London: Cambridge University Press.

Sölch, J. (1936). Geomorphologische probleme des Schottischen Hochlands. *Mitt. geogr. Geo. Wien,* **79.**

Steers, J. A. (1937). The Culbin Sands and Burghead Bay. *Geogr. Jl.,* **90,** 498–528.

Steven, H. M. and Carlisle, A. (1959). *The native pinewoods of Scotland.* Edinburgh and London: Oliver and Boyd.

Sugden, D. E. (1968). The selectivity of glacial erosion in the Cairngorm Mountains. *Trans. Inst. Br. Geogr.,* **45,** 79–92.

Synge, F. M. (1956). The glaciation of north-east Scotland. *Scott. geogr. Mag.,* **72,** 129–43.

The following soil maps deal with parts of the Eastern Scotland area:

SOIL MAPS

Glentworth, R., Hart, R., Dion, H. G., Muir, J. W., Laing, D., Shipley, B. M., Smith, J. and Grant, C. J. (1959). Soil map of Inverurie (Sheet 76). Scale 1:63 360. Chessington: Ordnance Survey.

Glentworth, R., Hart, R., Muir, J. W., Romans, J. C. C., Mitchell, B. D. and Mulcahy, M. J. (1954). Soil map of Huntly (Sheet 86). Scale 1:63 360. Chessington: Ordnance Survey.

Glentworth, R., Laing, D., Muir, J. W., Hart, R. and Mackenzie, R. C. (1962). Soil map of Aberdeen (Sheet 77). Scale 1: 63 360. Chessington: Ordnance Survey.

Glentworth, R., Mitchell, B. D. and Grant, R. (1954). Soil map of Banff (Sheet 96). Scale 1:63 360. Chessington: Ordnance Survey.

Glentworth, R., Muir, J. W., Romans, J. C. C., Birse, E. L., Smith, J. and Shipley, B. M. (1966). Soil map of Banchory and Stonehaven (Sheets 66 and 67). Scale 1:63 360. Southampton: Ordnance Survey.

Glentworth, R., Muir, J. W., Shipley, B. M., Grant, R., Bown, C. J., Hart, R. and Dion, H. G., (1962). Soil map of Peterhead and Fraserburgh (Sheets 87 and 97). Scale 1:63 360. Chessington: Ordnance Survey.

Grant, R., Birse, E. L. and Harper, P. C. (1956). Soil map of Elgin (Sheet 95). Scale 1:63 360. Chessington: Ordnance Survey.

Heslop, R. E. F. and Campbell, C. G. B. (1981). Soil map of Tomintoul (Sheet 75). Scale 1:63 360. Southampton: Ordnance Survey.

Laing, D., Romans, J. C. C., Lawrence, E., Walker, A. D., Bown, C. J, and Law, R. D. (1968). Soil map of Perth and Arbroath (Sheets 48 and 49). Scale 1:63 360. Southampton: Ordnance Survey.

Laing, D., Lawrence, E., Robertson, J. S. and Merrilees, D. W. (1975). Soil map of Kinross, Elie and Edinburgh (Sheet 40 and parts of Sheet 41 and 32). Scale 1:63 360. Southampton: Ordnance Survey.

Muir, J. W., Romans, J. C. C. Laing, D., Smith, J. and Glentworth, R. (1964). Soil map of Forfar (Sheets 57 and 57A). Scale 1:63 360. Southampton: Ordnance Survey.

Romans, J. C. C., Grant, R., Walker, A. D., Strachan, W. R. and Robertson, J. S. (1972). Soil map of Cromarty and Invergordon (Sheet 94). Scale 1:63 360. Southampton: Ordnance Survey.

Romans, J. C. C., Lang, D. M. and Cruickshank, J. (1972). Soil map of the Black Isle (Parts of Sheets 83, 84, 93, and 94). Scale 1:63 360. Southampton: Ordnance Survey.

Romans, J. C. C., Hudson, G., Grant, R., Birse, E. L. and Harper, P. C. (1980). Soil map of Rothes and Elgin (Sheets 85 and 95). Scale 1:63 360. Sheet 95 revision by R. E. F. Heslop and C. G. B. Campbell. Southampton: Ordnance Survey.

Shipley, B. M., Stevens, J. H., Lawrence, E. and Jarvis, R. A. (1968). Soil map of Stirling and part of Airdrie (Sheet 39 and part of 31). Scale 1:63 360. Southampton: Ordnance Survey.

Shipley, B. M., Stevens, J. H., Merrilees, D. W., Morris, R. J. F. and Wright, G. G. (1983). Soil map of Crieff (Sheet 47). Scale 1:63 360. Southampton: Ordnance Survey.

Walker, A. D., Grant, R., Law, R. D., Jack, J. I. and Gauld, J. H. (1976). Soil map of Nairn and Cromarty (Sheet 84 and part of 94). Scale 1:63 360. Southampton: Ordnance Survey.